Generis

PUBLISHING

I0047424

Deep Fundamentals and Wide Applications of Bayesian Inference Including Detailed Computer Codes

Hiroshi Isshiki

Title: Deep Fundamentals and Wide Applications of Bayesian Inference Including Detailed Computer Codes

ISBN: 979-8-88676-300-3

Author: Hiroshi Isshiki

Cover image: www.pixabay.com

Publisher: Generis Publishing
Online orders: www.generis-publishing.com
Contact email: info@generis-publishing.com

Table of Contents

1. INTRODUCTION

Readers who read table of contents of this book may have been surprised at the wide range of topics. Specifically, it includes the estimation of parameters of statistical distributions and differential equations, estimation of high-precision solutions by data assimilation, infection problems of new coronavirus, diagnosis of diseases, pattern recognition, and inference in the brain. The author thinks that all of them are suitable problems for treatment by Bayesian inference. Through this, you can understand the wide range of applications of Bayesian inference.

The author think that the method of explanation is also a feature of this book. The explanation method is not always smart, but the author thinks it is easy to understand. Since it is not always convenient for actual applications, it also explains the methods that should be adopted for applications. For example, when finding the maximum value of probability, the method of finding the maximum value by enumerating the probability values at multiple candidate points is described first. Next, the steepest ascent method, which is convenient for application, is described. Through such explanations, not only can we fully understand the problem without advanced mathematical knowledge, but we can naturally understand the existence of the local maximum values.

Since the advent of deep learning, the neural network has brought a big innovation in the world [1, 2, 3]. However, deep learning might be far from perfect, because of "the inference is a black box", "unexpected answer due to the overfitting", and "large scale of the network and long time learning". The earliest answer to them should be given. Among them, the black box nature would be a fundamental problem.

The Bayesian inference is based on a quite different theory as the neural network [4, 5, 6]. It might be free from a few problems of the neural networks. The Bayesian inference seems to have disappeared in the world of statistics for a while. This is because prior probabilities have a strong subjective aspect and have been repelled by some orthodox statisticians. However, when it comes to decision-making on issues such as management, politics, and society, it seems that judgments based on objective assumptions in every sense cannot be hoped for. In such a case, Bayesian inference that can incorporate subjectivity will be extremely practical.

In neural network learning, the difference between the neuron value of the output layer and the teacher data for a certain input is regarded as an error, and the weight is adjusted so that the error is minimized. That is, we have to solve the multivariable minimum value problem. On the other hand, in Bayesian learning using Bayesian inference, it means to obtain the probability distribution of the output conditional on the input from a large number of training data. In Bayesian learning, learning is to find the frequency distribution from the learning data. And, also in the case of Bayesian inference, we encounter the problem of the maximum value of multiple variables because we look for the one that maximizes the probability at the judgment stage, but the number of variables is much smaller than that of neural networks. In general, it will not raise a serious problem.

What is human memory? Neural networks are said to imitate the human brain [1, 2, 3]. The brain has an enormous number of brain cells, which are said to be in the hundreds of billions, and it is said that the brain cells are connected to each other by nerve fibers called axons to form a network. This network has the function of transmitting and controlling information between brain cells, and what kind of information processing is performed is formed by learning. Neural networks are mathematical models of this hypothesis. Therefore, the memory is stored in the network.

There is another way of thinking. It is called the grandmother hypothesis, and it is a hypothesis that there is a cell that reacts specifically when you look at the grandmother. This means that memory is stored in a single cell or multiple cells, which is very different from the network theory described above. For example, the handwritten character "1" is formed in brain cells by looking at various "1" s. The frequency distribution (probability distribution) is made in a grid pattern for the character "1". It can be considered that the grid pattern for the character "1" has become established in the cell set in this way. This kind of thinking arises from the mathematical model of number pattern recognition by Bayesian inference discussed in chapter 7.

In this way, Bayesian inference is based on learning and inference that are completely different from neural networks [4, 5, 6]. It may be unrelated to some of the problems with neural networks.

Bayesian inference performs inference similar to neural networks. Bayesian inference, like neural networks, learns and infers based on data, but it is more deductive and less data-dependent than neural networks. The difference appears as a difference in the number of unknown variables. Neural networks have a much larger number of unknown variables. Although it can be said that the flexibility is high, overlearning is likely to occur and the basis of reasoning becomes incomprehensible.

On the other hand, the learning of Bayesian inference is to find the likelihood function from the frequency distribution, and to find the posterior probability from the likelihood function and the prior distribution. Therefore, it can be said that the learning and reasoning of Bayesian inference are highly deductive. However, it is powerless when the relationship between cause and effect cannot be clearly mathematically modeled, such as when discriminating figures. Humans can easily discriminate figures and understand languages. However, it is extremely difficult to clearly mathematically model the process of judgment and understanding for these. Human judgment and understanding in this regard is a black box.

Neural networks make learning decisions while remaining in the black box. In other words, neural networks are extremely flexible because they can be learned simply by giving input data and teacher data. In comparison, Bayesian inference requires some modeling, so it must be said that it is less flexible. From another point of view, it can be said that the mechanism of neural networks is closer to human learning and reasoning than to Bayesian reasoning. The price is that human learning, like neural network learning, takes a lot of time.

However, in the case of Bayesian inference, the problem of the maximum value of multiple variables appears at the judgment stage for finding the maximum probability. Generally speaking, it is not a huge number of variables such as network weights encountered in the learning stage of neural networks, but a maximum value problem of a relatively small number of variables.

Numerical simulation technology has made remarkable progress. Calculation accuracy, speed, and scale of the target have improved dramatically. Complex fluid motion, combustion problems so on can be calculated using state-of-the-art supercomputers. On the other hand, the limits of numerical simulation have become clear. If the object is not captured in a completely explicit form, problems such as earthquake prediction, for example, could not be dealt with satisfactorily. Even in such a case, the situation will be different if there is something to supplement the unknown part, for example, observation data. Earthquake prediction is still difficult, but depending on the problem, by combining observation data with numerical simulation, it would be possible to grasp the phenomenon with considerable accuracy. The effectiveness of combining data observation and numerical simulation was recognized, and it came to be called data assimilation technology.

It is a technology for learning using data, but the difference from neural networks is that it uses a mathematical model that expresses phenomena as its skeleton. A neural network is a method of learning using only data without a model. Since data assimilation is a method of learning the unknown part of a mathematical model with data, it enables efficient and highly accurate inference. In data assimilation, knowledge of phenomena and their mathematics and knowledge of statistics that are the basis of data assimilation, especially Bayesian statistics, are required. We hope that the explanations in this book will help beginners understanding.

In this book, much numerical application of Bayesian inference is discussed to help the reader understand how to use Bayesian inference concretely. Furthermore, a rather unfamiliar application such as reasoning in the human brain is also introduced.

The reasoning in our brain is also based on Bayes inference. Friston's principle of minimum free energy [6] has made it possible to interpret the function of the cerebrum in a unified manner. The principle is conceptually and mathematically novel and sophisticated, so it is difficult to understand it accurately. The principle is explained from a slightly different angle, and some examples such as simple control are shown through numerical calculation.

We discuss the basic aspect of the Bayesian inference in chapter 2, the parameter identification of probability distribution in chapter 3, interpolation of function and identification of differential equation in chapter 4, data assimilation in chapter 5, application of data assimilation to infection of disease in chapter 6, diagnosis of illness based on the checklist in chapter 7, patter recognition in chapter 8, and the minimum principle of free energy for the reasoning in the human brain in chapter 9.

REFERENCES 1

[1] A. Krizhevsky, I. SutskeverI, and G. Hinton. "ImageNet classification with deep convolutional neural networks," In Proc. Advances in Neural Information Processing Systems 25 1090–1098 2012.

[2] Yann LeCun1, Yoshua Bengio & Geoffrey Hinton, "Deep learning," NATURE | VOL 521 | 28 MAY 2015.

[3] Ian Goodfellow, Yoshua Bengio and Aaron Courville, Deep Learning, The MIT Press 2017.

[4] A. Wald, Statistical Decision Functions, John Wiley & Sons, Inc. 1956.

[5] M. Antonia Amaral Turkman, Carlos Daniel Paulino, Peter Muller, Computational Bayesian Statistics, Cambridge Univ. Press, 2019.

[6] K. Friston, Is the free-energy principle neurocentric? Nature Reviews | Neuroscie, Vol. 11, 2010.

2. WHAT IS BAYESIAN INFERENCE?

2.1. Probabilistic estimation of medical test

Let the probability of P(*Result* | *Cause*) of *Result* under *Cause* be given. The reverse probability P(*Cause* | *Result*) of the cause that brought the result is obtained by the Bayesian theorem. The Bayesian inference estimates the reverse probability using the Bayesian theorem.

The Bayesian theorem is given by

$$P(Cause \mid Result) = \frac{P(Result, Cause)}{P(Result)} = \frac{P(Result \mid Cause)P(Cause)}{P(Result)}. \tag{2.1}$$

This is merely a mathematical theorem. *P*(*Cause*), *P*(*Result* | *Cause*), and P(*Cause* | *Result*) are called the prior probability, likelihood function, and posterior probability, respectively. In the following, we show concretely how the Bayesian theorem is used.

Let 5% of a population be patients *Pat* of disease, and 95% be healthy people *Nor*. They are called the prior probabilities, namely

$$P(Pat) = 0.05, \quad P(Nor) = 0.95. \tag{2.2}$$

The Bayesian inference was rejected for a while because of the prior probabilities. They do not cause any problem when the prior probabilities are given objectively. In some cases such as business, political and social problems, we are frequently obliged to give them subjectively. However, this property of the Bayesian inference has now become the characteristics of the Bayesian inference.

According to a test, 98% of the patients are positive (*Ptv*) and 2% of those negative (*Ntv*). On the other hand, 4% and 96% of the healthy people are positive and negative, respectively. Namely

$$\begin{aligned} &P(Ptv \mid Pat) = 0.98, \quad P(Ntv \mid Pat) = 0.02, \\ &P(Ptv \mid Nor) = 0.04, \quad P(Ntv \mid Nor) = 0.96. \end{aligned} \tag{2.3}$$

We assume that a person who is not identified as patient or non-patient is positive to the test. The probability that a person is a patient is estimated as follows using the Bayesian theorem. In the Bayesian inference, the learning is nothing but to obtain the probabilities.

Since we have the joint probabilities:

$$\begin{aligned} P(Ptv, Pat) &= P(Ptv \mid Pat)P(Pat) = 0.98 \times 0.05 = 0.049 \\ P(Ptv, Nor) &= P(Ptv \mid Nor)P(Nor) = 0.04 \times 0.95 = 0.038 \end{aligned} \tag{2.4}$$

from equations (2.2) and (2.3), we obtain the probabilities of being a patient or non-patient among people who are positive to the test:

$$P(Pat \mid Ptv) = \frac{P(Ptv, Pat)}{P(Ptv, Pat) + P(Ptv, Nor)} = \frac{0.049}{0.087} = 0.563,$$
$$P(Nor \mid Ptv) = \frac{P(Ptv, Nor)}{P(Ptv, Pat) + P(Ptv, Nor)} = \frac{0.038}{0.087} = 0.437. \tag{2.5}$$

Hence, the probability that the person is a patient is 56.3% and a non-patient 43.7%.

Since we have similarly, if the test result is negative

$$P(Ntv, Pat) = P(Ntv \mid Pat)P(Pat) = 0.02 \times 0.05 = 0.001$$
$$P(Ntv, Nor) = P(Ntv \mid Nor)P(Nor) = 0.96 \times 0.95 = 0.912, \tag{2.6}$$

we obtain the probabilities of being a patient or non-patient among people who are negative to the test:

$$P(Pat \mid Ntv) = \frac{P(Ntv, Pat)}{P(Ntv, Pat) + P(Ntv, Nor)} = \frac{0.001}{0.913} = 0.001$$
$$P(Nor \mid Ntv) = \frac{P(Ntv, Nor)}{P(Ntv, Pat) + P(Ntv, Nor)} = \frac{0.912}{0.913} = 0.999. \tag{2.7}$$

The probability that the person is a patient is 0.1% and a non-patient 99.9%.

We call $P(Nor|Pos) = 0.437$ and $P(Pat|Ntv) = 0.001$ as false positive and false negative, respectively. There is a case where false positives and false negatives can't be non-negligible.

Rewriting (2.5), we have

$$P(Pat \mid Ptv) = \frac{P(Ptv, Pat)}{P(Ptv, Pat) + P(Ptv, Nor)}$$
$$= \frac{P(Ptv \mid Pat)P(Pat)}{P(Ptv \mid Pat)P(Pat) + P(Ptv \mid Nor)P(Nor)}$$
$$P(Nor \mid Ptv) = \frac{P(Ptv, Nor)}{P(Ptv, Pat) + P(Ptv, Nor)}$$
$$= \frac{P(Ptv \mid Nor)P(Nor)}{P(Ptv \mid Pat)P(Pat) + P(Ptv \mid Nor)P(Nor)}. \tag{2.8}$$

Hence, if the prior probabilities are equal, namely

$$P(Pat) = P(Nor), \tag{2.9}$$

we obtain

$$P(Pat \mid Ptv) = \frac{P(Ptv \mid Pat)}{P(Ptv \mid Pat) + P(Ptv \mid Nor)} = \frac{P(Ptv \mid Pat)}{P(Ptv)}$$
$$P(Nor \mid Ptv) = \frac{P(Ptv \mid Nor)}{P(Ptv \mid Pat) + P(Ptv \mid Nor)} = \frac{P(Ptv \mid Nor)}{P(Ptv)}. \quad (2.10)$$

In this case, we can use the likelihood functions instead of the posterior probabilities for inference. It is nothing but the likelihood estimation.

2.2. Judgment of infection

2.2.1. Is the person who coughs infected with the new coronavirus?

In a country with a population of 100 million, we consider there are 2,000 people infected with the new coronavirus (hereinafter referred to as Corona) and 99,998,000 non-infected people with the new coronavirus (hereinafter referred to as NCorona). At this time, the probabilities of being a corona-infected person and a non-corona-infected person are given by the following equations, respectively:

$$P(Corona) = 2{,}000/100{,}000{,}000 = 0.00002, \quad (2.11a)$$

$$P(NCorona) = 99{,}998{,}000/100{,}000{,}000 = 0.99998. \quad (2.11b)$$

As data for people who cough, we assume the conditional probability of those who cough among corona-infected persons and the conditional probability of those who cough among non-corona-infected persons as follows:

$$P(Cough \mid Corona) = 1.0, \quad (2.12a)$$

$$P(Cough \mid NCorona) = 0.1. \quad (2.12b)$$

From equations (2.11) and (2.12), the joint probability of a corona-infected person and coughing person and the non-corona-infected person and coughing person are given by

$$\begin{aligned} P(Corona, Cough) &= P(Cough \mid Corona)P(Corona) \\ &= 1.0 \times 0.00002 = 0.00002 \end{aligned}, \quad (2.13a)$$

$$\begin{aligned} P(NCorona, Cough) &= P(Cough \mid NCorona)P(NCorona) \\ &= 0.1 \times 0.99998 = 0.099998 \end{aligned}, \quad (2.13b)$$

Therefore, according to Bayes' theorem, the conditional probability that a person who coughs is a corona-infected person and the conditional probability that a person who coughs is a non-corona-infected person are obtained as

$$P(Corona \mid Cough) = \frac{P(Corona, Cough)}{P(Corona, Cough) + P(NCorona, Cough)}$$
$$= \frac{0.00002}{0.00002 + 0.099998} = 0.00020, \tag{2.14a}$$

$$P(NCorona \mid Cough) = \frac{P(NCorona, Cough)}{P(Corona, Cough) + P(NCorona, Cough)}$$
$$= \frac{0.099998}{0.00002 + 0.099998} = 0.99980. \tag{2.14b}$$

It can be seen that the conditional probability of a person who coughs among corona-infected persons is 1.0, but the conditional probability that a person who coughs is a corona-infected person is 0.0002, which is extremely low.

Table 2.1 shows the results calculated by changing the conditional probabilities of people who cough among non-corona-infected persons while keeping the conditional probability of those who cough among corona-infected persons as 1.

Table 2.1 Probability of a coughing person having a new coronavirus infection

ᵖ(Cough|NCorona)	P(Corona|Cough)	P(NCorona|Cough)
0/10000	1.000000	0
1/10000	0.001996	0.998004
2/10000	0.000999	0.999001
3/10000	0.000666	0.999334
4/10000	0.000500	0.999500
5/10000	0.000400	0.999600
6/10000	0.000333	0.999667
7/10000	0.000286	0.999714
8/10000	0.000250	0.999750
9/10000	0.000222	0.999778
10/10000	0.000200	0.999800

Even though a corona patient is a 100% coughing person, a person who happens to be coughing is not necessarily a corona patient. Assuming that 10% of non-corona patients cough, the probability that a person who happens to be a corona patient is only 0.02%! You don't have to worry too much.

2.2.2. If you have a high fever, are you infected with the new coronavirus?

In a country with a population of 100 million, we consider there are 2,000 people infected with the new coronavirus (hereinafter referred to as Corona) and

99,998,000 non-infected people with the new coronavirus (hereinafter referred to as NCorona). At this time, the probabilities of being a corona-infected person and a non-corona-infected person are given by the following equations, respectively:

$$P(Corona) = 2,000/100,000,000 = 0.00002, \tag{2.15a}$$

$$P(NCorona) = 99,998,000/100,000,000 = 0.99998. \tag{2.15b}$$

As data on people with high fever, the conditional probability of those with high fever among corona-infected persons and the conditional probability of those with high fever among non-corona-infected persons are assumed as

$$P(HTmp \mid Corona) = 1.0, \tag{2.16a}$$

$$P(HTmp \mid NCorona) = 0.01. \tag{2.16b}$$

From equations (2.15) and (2.16), the joint probabilities of corona-infected persons with high fever and non-corona-infected persons with high fever are obtained as

$$\begin{aligned} &P(Corona, HTmp) \\ &\quad = P(HTmp \mid Corona)P(Corona) = 1.0 \times 0.00002 = 0.00002, \end{aligned} \tag{2.17a}$$

$$\begin{aligned} &P(NCorona, HTmp) \\ &\quad = P(HTmp \mid NCorona)P(NCorona) = 0.01 \times 0.99998 = 0.0099998. \end{aligned} \tag{2.17b}$$

Therefore, according to Bayes' theorem, the conditional probability that a person with a high fever is a corona-infected person and the conditional probability that a person with a high fever is a non-corona-infected person are given by

$$\begin{aligned} P(Corona \mid HTmp) &= \frac{P(Corona, HTmp)}{P(Corona, HTmp) + P(NCorona, HTmp)} \\ &= \frac{0.00002}{0.00002 + 0.0099998} = 0.001996, \end{aligned} \tag{2. 18a}$$

$$\begin{aligned} P(NCorona \mid HTmp) &= \frac{P(NCorona, HTmp)}{P(Corona, HTmp) + P(NCorona, HTmp)} \\ &= \frac{0.0099998}{0.00002 + 0.0099998} = 0.99800. \end{aligned} \tag{2. 18b}$$

Although the conditional probability of a person with high fever is 1.0 among those infected with corona, the conditional probability that a person with high fever is infected with corona is 0.002, which is extremely low.

Table 2.2 shows the results of calculations with various changes in the conditional probabilities of people with high fever among non-corona-infected persons, while keeping the conditional probability of those with high fever of 1 among those infected with corona.

Table 2.2 Probability of a person with a high fever having a new coronavirus infection

P(HTemp|NCorona)	'(Corona| HTemp)	P(NCorona| HTemp)
0/1000	**0**	1
10/1000	0.00200	0.99800
20/1000	0.00100	0.99900
30/1000	0.00067	0.99933
40/1000	0.00050	0.99950
50/1000	0.00040	0.99960
60/1000	0.00033	0.99967
70/1000	0.00029	0.99971
80/1000	0.00025	0.99975
90/1000	0.00022	0.99978
100/1000	0.00020	0.99980

Even though a corona patient has a 100% high fever, a person who happens to have a high fever is not necessarily a corona patient. Assuming that 10% of non-corona patients have a high fever, the probability that a person who happens to be a corona patient is only 0.02%! You don't have to worry too much.

2.2.3. If you have a cough and a high fever, are you infected with the new coronavirus?

Consider the case where there are multiple symptoms. For example, it is not just a high fever, but a cough and a high fever.

In the following calculation, equation (2.12) is assumed for cough and equation (2.16) is assumed for fever as symptom data. At this time, the probabilities $P(Corona \mid Cough)$ of a coughing corona-infected person and $P(NCorona \mid Cough)$ of a coughing corona-non-infected person are calculated as Eq. (2.14), respectively:

$$P(Corona \mid Cough) = 0.00020, \tag{2.19a}$$

$$P(NCorona \mid Cough) = 0.99980 \tag{2.19b}$$

This can be used as a prior probability for Bayesian inference of a person who is coughing and has a higher fever (see equation (2A.1) in the appendix).

As data on people with high fever, the conditional probabilities $P(HTmp \mid Corona)$ of those with high fever among corona-infected persons and the conditional probabilities $P(HTmp \mid NCorona)$ of those with high fever among non-corona-infected persons are as follows from Eq. (2.16):

$$P(HTmp \mid Corona) = 1.0, \tag{2.20a}$$
$$P(HTmp \mid NCorona) = 0.01 \tag{2.20b}$$

From equations (2.19) and (2.20), the joint probability $P(Corona, CghHT)$ of a corona-infected person with cough and high fever and the joint probability $P(NCorona, CghHT)$ of a non-corona-infected person with cough and high fever are obtained as (see equation (2A.1) in the appendix):

$$\begin{aligned} &P(Corona, CghHT) \sim P(HTmp \mid Corona)P(Corona \mid Cough) \\ &= 1.0 \times 0.00020 = 0.00020 \end{aligned}, \tag{2.21a}$$

$$\begin{aligned} &P(NCorona, CghHT) \sim P(HTmp \mid NCorona)P(NCorona \mid Cough) \\ &= 0.01 \times 0.99980 = 0.0099800 \end{aligned} \tag{2.21b}$$

Therefore, according to Bayes' theorem, the conditional probability $P(Corona \mid CghHT)$ that a person with cough and high fever is a corona-infected person and the conditional probability $P(NCorona \mid CghHT)$ that a person with cough and high fever is a non-corona-infected person are given by

$$\begin{aligned} P(Corona \mid CghHT) &= \frac{P(Corona, CghHT)}{P(Corona, CghHT) + P(NCorona, CghHT)} \\ &= \frac{0.00020}{0.00020 + 0.00998} = 0.019646, \end{aligned} \tag{2.22a}$$

$$\begin{aligned} P(NCorona \mid CghHT) &= \frac{P(NCorona, CghHT)}{P(Corona, CghHT) + P(NCorona, CghHT)} \\ &= \frac{0.00998}{0.00020 + 0.00998} = 0.980354 \end{aligned} \tag{2.22b}$$

It can be seen that the conditional probability $P(Corona \mid CghHT)$ that a person with cough and high fever is a corona-infected person is 0.0196, which is considerably higher than 0.00020 of $P(Corona \mid Cough)$ and 0.0020 of $P(Corona \mid HTtmp)$. On the other hand, it can be seen that the conditional probability $P(NCorona \mid CghHT)$ that a person with cough and high fever is a non-corona infected person is 0.9803, which is slightly smaller than 0.9998 of $P(NCorona \mid Cough)$ and 0.9980 of $P(NCorona \mid HTtmp)$. However, it still follows:

$$P(Corona \mid CghHT) \ll P(NCorona \mid CghHT) \tag{2.23}$$

People who have a cough and high fever are considered to be non-infected with corona.

2.3. Estimation of nationality

Given that there are distribution probabilities for height, weight, eye color, hair color, and skin color for each nationality, consider how to estimate nationality using Bayesian inference from these characteristics. For the sake of simplicity, we will limit ourself to two characteristics: height and weight.

In Bayesian inference, the following procedure is followed.

(1) Prior probability ... Probability of being a national

(2) Data probability (likelihood function) *P* (*height classification, weight classification* | *nationallity*) ... Data probability (likelihood function) *P* (*height classification, weight classification* | *nationality*) ... Height and weight probability distribution of each country)

(3) Data probability × prior probability → joint probability → P (nationality | height classification, weight classification) ... For example, P (American | 180cm-200cm, 70kg-80kg)

Figure 2.1 shows an image of the height distribution probability.

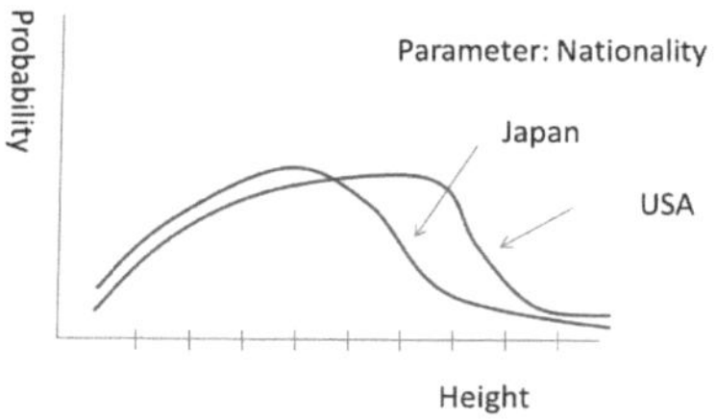

Fig. 2.1 Probability distribution of height.

Assuming that there is a distribution probability of height and weight for each nationality, a specific calculation method for estimating nationality from height and weight is shown below. First, we estimate the prior probabilities $P(Nat_n)$ of nationality Nat_n. The data probability (likelihood function) $P(Hgt_i, Wgt_j|Nat_n)$ of height Hgt_i and weight Wgt_j under the condition of nationality is supposed to be known approximately:

$$P(Hgt_i, Wgt_j \mid Nat_n) = P(Hgt_i \mid Nat_n)P(Wgt_j \mid Nat_n). \qquad (2.24)$$

The joint probabilities of height, weight, and nationality are calculated from the prior probabilities of nationalities and the conditional probabilities given by equation (2.24):

$$P(Hgt_i, Wgt_j, Nat_n) = P(Hgt_i, Wgt_j \mid Nat_n)P(Nat_n). \qquad (2.24)$$

Therefore, the inverse probability $P(Nat_n|Hgt_i, Wgt_j)$ is calculated as

$$P(Nat_n \mid Hgt_i, Wgt_j) = \frac{P(Hgt_i, Wgt_j, Nat_n)}{P(Hgt_i, Wgt_j)} = \frac{P(Hgt_i, Wgt_j, Nat_n)}{\sum_n P(Hgt_i, Wgt_j, Nat_n)}. \quad (2.25)$$

Appendix 2A Calculation of Bayesian inference

(1) When a person has a cough and a high fever

The following equation holds for a corona-infected person $P(Corona, CghHT)$ who has a cough and a high fever and a non-corona-infected person $P(NCorona, CghHT)$ who has a cough and a high fever:

$$\begin{aligned} P(Corona, CghHT) &= P(HTmp \mid Corona)P(Cough \mid Corona)P(Corona) \\ &= P(HTmp \mid Corona)P(Corona \mid Cough)P(Cough) \\ &\sim P(HTmp \mid Corona)P(Corona \mid Cough), \end{aligned} \quad (2A.1)$$

$$\begin{aligned} P(NCorona, CghHT) &= P(HTmp \mid NCorona)P(Cough \mid NCorona)P(NCorona) \\ &= P(HTmp \mid NCorona)P(NCorona \mid Cough)P(Cough) \\ &\sim P(HTmp \mid NCorona)P(NCorona \mid Cough). \end{aligned}$$

(2A.2)

The following approximate equation is used in equation (2A.1).

$$P(CghHT \mid Corona) = P(Cough \mid Corona)P(HTmp \mid Corona). \quad (2A.3)$$

The same applies to equation (2A.2).

Therefore, we have

$$\begin{aligned} P(Corona \mid CghHT) &= \frac{P(Corona, CghHT)}{P(Corona, CghHT) + P(NCorona, CghHT)} \\ &= \frac{P(HTmp \mid Corona)P(Corona \mid Cough)}{P(HTmp \mid Corona)P(Corona \mid Cough) + P(HTmp \mid NCorona)P(NCorona \mid Cough)}, \end{aligned}$$

(2A.4)

$$\begin{aligned} P(NCorona \mid CghHT) &= \frac{P(NCorona, CghHT)}{P(Corona, CghHT) + P(NCorona, CghHT)} \\ &= \frac{P(HTmp \mid NCorona)P(NCorona \mid Cough)}{P(HTmp \mid Corona)P(Corona \mid Cough) + P(HTmp \mid NCorona)P(NCorona \mid Cough)}. \end{aligned}$$

(2A.5)

In other words, it can be seen that if the prior probabilities $P(Corona)$ and $P(NCorona)$ are changed to $P(Corona \mid Cough)$ and $P(NCorona \mid Cough)$ as the result of the first Bayesian estimation in the second Bayesian inference, the Bayesian inference when coughing and high fever occurrence can be performed.

(2) When a person has a high fever for two days in a row

Regarding corona-infected person $P(Corona, 2HTmp)$ who had a high fever for two days and non-corona-infected person $P(NCorona, 2HTmp)$ who had a high fever for two days, we have

$$\begin{aligned}P(Corona, 2HTmp) &= P(HTmp \mid Corona)P(HTmp \mid Corona)P(Corona)\\ &= P(HTmp \mid Corona)P(Corona \mid HTmp)P(HTmp) \\ &\sim P(HTmp \mid Corona)P(Corona \mid HTmp),\end{aligned} \tag{2A.6}$$

$$\begin{aligned}P(NCorona, 2HTmp) &= P(HTmp \mid NCorona)P(HTmp \mid NCorona)P(NCorona)\\ &= P(HTmp \mid NCorona)P(NCorona \mid HTmp)P(HTmp)\\ &\sim P(HTmp \mid NCorona)P(NCorona \mid HTmp).\end{aligned} \tag{2A.7}$$

Strictly speaking, the following equation holds approximaely:

$$P(2HTmp \mid Corona) = P(HTmp \mid Corona)P(HTmp \mid Corona) \tag{2A. 8}$$

It can be considered effective as an approximation.

Therefore, we have

$$\begin{aligned}P(Corona \mid 2HTmp) &= \frac{P(Corona, 2HTmp)}{P(Corona, 2HTmp) + P(NCorona, 2HTmp)}\\ &= \frac{P(HTmp \mid Corona)P(Corona \mid HTmp)}{P(HTmp \mid Corona)P(Corona \mid HTmp) + P(HTmp \mid NCorona)P(NCorona \mid HTmp)},\end{aligned} \tag{2A.9}$$

$$\begin{aligned}P(NCorona \mid 2HTmp) &= \frac{P(NCorona, 2HTmp)}{P(Corona, 2HTmp) + P(NCorona, 2HTmp)}\\ &= \frac{P(HTmp \mid NCorona)P(NCorona \mid HTmp)}{P(HTmp \mid Corona)P(Corona \mid HTmp) + P(HTmp \mid NCorona)P(NCorona \mid HTmp)}.\end{aligned} \tag{2A.10}$$

In other words, if the prior probabilities $P(Corona)$ and $P(NCorona)$ are changed to $P(Corona \mid HTmp)$ and $P(NCorona \mid HTmp)$ as the result of the first Bayesian estimation in the second Bayesian inference, it can be seen that the Bayesian inference can be performed when the fever is generated for two consecutive days.

3. ESTIMATION OF PARAMETERS OF PROBABILITY DISTRIBUTION

First, the basic idea of this chapter is described. The purpose of this chapter is to use Bayesian inference to estimate the parameters of the probability distribution that produced the data, such as the mean and variance of a normal distribution, from a series of data generated by a probability distribution. In Bayesian inference in this case, the conditional probability $P(x \mid \mu)$ in which a series of data x is generated for a known parameter μ is called a likelihood function. In the reverse process, the conditional probability $P(\mu \mid x)$ for estimating the unknown parameter is obtained by assuming that a series of data is known and the parameter that generated the data is unknown. The latter probability is called the posterior probability.

Bayes' inference uses Bayes' theorem in Eq. (2.1). When applied to parameter estimation, Bayes' theorem is given by:

$$P(\mu \mid x) \equiv \frac{P(x,\mu)}{P(x)} = \frac{P(x \mid \mu)P(\mu)}{P(x)}.$$

In this chapter, Bayesian inference uses the likelihood method. In other words, we consider various candidates for μ, and consider that the one that maximizes the posterior probability $P(\mu \mid x)$ or likelihood function $P(x \mid \mu)$ gives the correct parameter μ. The former maximizing method is called MAP (or maximum a posteriori) method, and the latter maximum likelihood method.

There are various possible methods for searching for the maximum value, but the simplest and easiest to understand is to list multiple discrete value candidates and select from them. In this chapter, this method is mainly used with an emphasis on comprehensibility. For continuous variables, the Steepest Ascent method is used generally. Since the probability takes a value between 0 and 1, the value of the likelihood function becomes very small and may cause underflow. In such a case, it is effective to take the natural logarithm. Even if the logarithm is taken, the relationship of magnitude does not change.

3.1. Bernoulli distribution

Let x=1 and x=0 refer to the face and back in coin throw, respectively. The probability $P(x)$ is called Bernoulli distribution and given by

$$P(x \mid \mu) = \mu^{x}(1-\mu)^{1-x}. \tag{3.1}$$

The parameter μ is the probability of x=1. We infer the parameter μ when a sequence of random numbers $\boldsymbol{x}=x_1, x_2, \ldots, x_N$ consisting of 1 and 0 is given.

According to the Bayesian theorem, the reverse probability $P(\mu \mid \mathbf{x})$ is given by

$$P(\mu \mid \mathbf{x}) = \frac{P(\mathbf{x}, \mu)}{P(\mathbf{x})} = \frac{P(\mathbf{x} \mid \mu) P(\mu)}{P(\mathbf{x})}. \tag{3.2}$$

When the number of the candidates of the parameter μ is $\mu_1, \mu_2, \ldots, \mu_I$, the probability of $\mu=\mu_i$ can be obtained by

$$P(\mu_i \mid \mathbf{x}) = \frac{P(\mathbf{x} \mid \mu_i) P(\mu_i)}{\sum_{j=1}^{I} P(\mathbf{x}, \mu_j)} = \frac{P(\mathbf{x} \mid \mu_i) P(\mu_i)}{\sum_{j=1}^{I} P(\mathbf{x} \mid \mu_j) P(\mu_j)}. \tag{3.3}$$

If we assume

$$P(\mu_1) = P(\mu_2) = \cdots = P(\mu_I), \tag{3.4}$$

equation (3.3) becomes

$$P(\mu_i \mid \mathbf{x}) = \frac{P(\mathbf{x} \mid \mu_i)}{\sum_{j=1}^{I} P(\mathbf{x} \mid \mu_j)} \sim P(\mathbf{x} \mid \mu_i), \tag{3.5}$$

since the denominator dose not include μ_i and does bot contribute in the search of the maximum. This is nothing but the likelihood method. The appropriateness is discussed later.

The likelihood function $P(\mathbf{x} \mid \mu_i)$ could be calculated by

$$P(\mathbf{x} \mid \mu_i) = \prod_{n=1}^{N} P(x_n \mid \mu_i). \tag{3.6}$$

The μ_i that makes $P(\mu_i \mid \mathbf{x})$ given by equation (3.5) the maximum gives the estimation of the parameter μ. This is nothing but the estimation by the maximum likelihood method. The continuous search of unknown parameters are explained in Appendix 3A.

Numerical calculations were conducted. The programming code is shown in Appendix 3B. A numerical example of the above-mentioned estimation method is shown below. Let $\boldsymbol{x}=x_1, x_2, \ldots, x_N$ be a random sequence of 0 and 1 with N=100 generated by Bernoulli distribution given by equation (3.1) with μ=0.35. The random sequence is shown in Fig. 3.1.

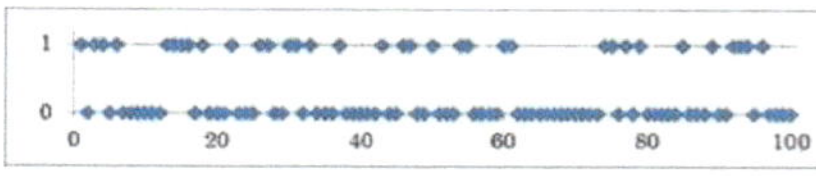

Fig. 3.1 A random sequence that follows Bernoulli distribution

The calculation result is shown in Fig. 3.2, where the candidates of μ are given by

$$.\mu_i = \frac{i}{10}, \quad i = 0, 1, \cdots, 10 \tag{3.7}$$

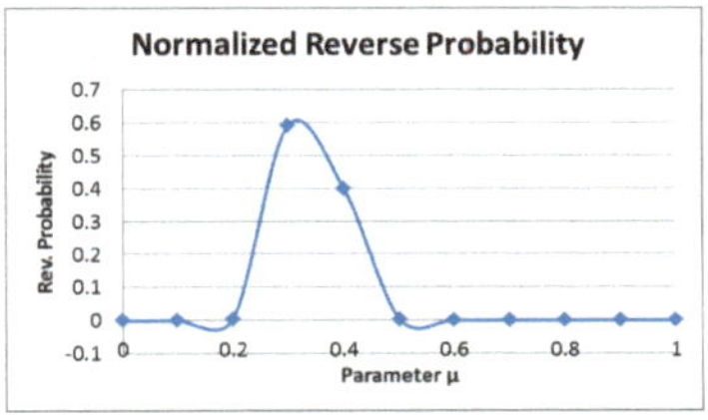

Fig. 3.2 The reverse probability (The candidates of μ are coarsely set)

If we use the finer setting for the candidates as

$$\mu_i = \frac{i}{20}, \quad i = 0, 1, \cdots, 20, \tag{3.8}$$

we have a result as shown in Fig. 3.3. Since μ=0.35 gives the maximum value, this value could be the estimation.

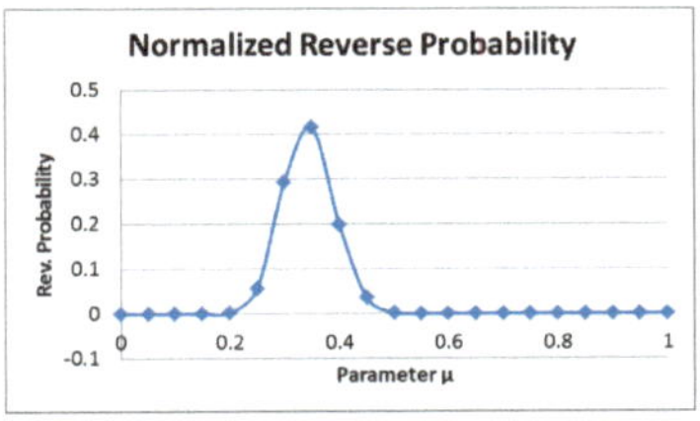

Fig. 3.3 The reverse probability (The candidates of μ are finely set).

For reference, the results of N=50 and N=200 are shown in Fig. 3.4 (a) and Fig. 3.4 (b), respectively. In the case of N=50, the estimation of μ=0.4 is not accurate. In the case of N=200, the result is correct and the peak becomes sharper.

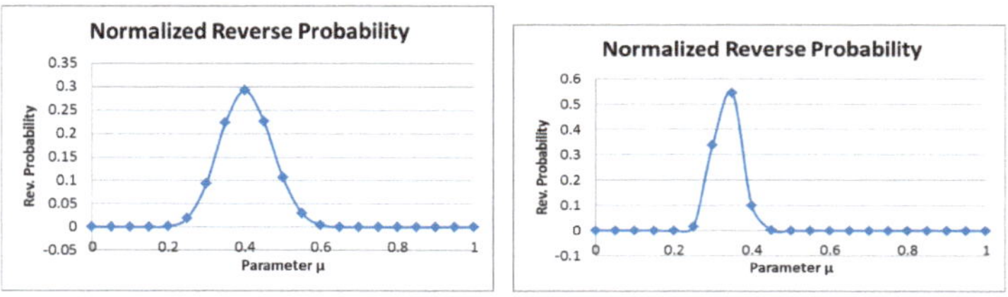

(a) N=50 (b) N=200

Fig. 3.4 Effects of the length of the random sequence N on the reverse probability.

If the prior probabilities are equal, the Bayes inference is equal to the maximum likelihood method. If the prior probabilities are not equal, a result different from that of the maximum likelihood method might be obtained. However, in the present case, if we increase the number of data N, the result does not depend on the choice of the prior probabilities. We show this property using numerical examples below.

If the prior probability is given by a normal distribution with the average μ=0.5 and the standard deviation σ=0.25:

$$P(\mu_i) = \frac{1}{\sqrt{2\pi\sigma^2}} \exp\left(-\frac{(\mu_i - \mu)^2}{2\sigma^2} \right), \tag{3.9}$$

the parameter μ_i estimated using (3.3) converges to 0.35 as N increases.

Table 3.1 The Bayesian estimates of μ_i with the unequal choices of the prior probabilities

Number of Data N	Estimated μ_i	Max of $P(\mu_i)$
25	0.4	0.21
50	0.4	0.292
100	0.35	0.415
200	0.35	0.543
400	0.35	0.684

If we assume that the prior probability is given by a normal distribution with the average μ=0.75 and the standard deviation σ=0.125, we also obtain the same result. This means that the prior probabilities do not affect the estimate as far as we use as big N as sufficient. This property originates from the fact that the data are generated from a single source. If the data are generated from several sources, the prior probabilities possibly affect the posterior probabilities.

3.2. Normal distribution

We assume two parameters of a normal distribution are the mean μ=0 and standard deviation σ=1. Let the candidate of μ and σ be discretized as

$$\begin{aligned} \mu_i &= -3.0 + 0.25i, \quad i = 0,1,\cdots,24 \\ \sigma_j &= 0.125 + 0.125j, \quad j = 0,1,\cdots,24 \end{aligned} \tag{3.10}$$

Equation (3.3) in case of Bernoulli distribution is replaced by

$$P(\mu_i,\sigma_j \mid \mathbf{x}) = \frac{P(\mathbf{x} \mid \mu_i,\sigma_j)P(\mu_i)P(\sigma_j)}{\sum_{i=1}^{I}\sum_{j=1}^{J} P(\mathbf{x},\mu_i,\sigma_j)} = \frac{P(\mathbf{x} \mid \mu_i,\sigma_j)P(\mu_i)P(\sigma_j)}{\sum_{i=1}^{I}\sum_{j=1}^{J} P(\mathbf{x} \mid \mu_i,\sigma_j)P(\mu_i)P(\sigma_j)}. \tag{3.11}$$

With respect to the prior probabilities, if we assume

$$\begin{aligned} P(\mu_1) &= P(\mu_2) = \cdots = P(\mu_I) \\ P(\sigma_1) &= P(\sigma_2) = \cdots = P(\sigma_I) \end{aligned}, \tag{3.12}$$

(3.11) becomes

$$P(\mu_i,\sigma_j \mid \mathbf{x}) = \frac{P(\mathbf{x} \mid \mu_i,\sigma_j)}{\sum_{i=1}^{I}\sum_{j=1}^{J} P(\mathbf{x} \mid \mu_i,\sigma_j)} \sim P(\mathbf{x} \mid \mu_i,\sigma_j), \tag{3.13}$$

Since μ_i and σ_j do not contribute the maximization. This is nothing but the likelihood method.

The likelihood function $P(\mathbf{x} \mid \mu_i, \sigma_j)$ could be calculated by

$$P(\mathbf{x} \mid \mu_i,\sigma_j) = \prod_{n=1}^{N} P(x_n \mid \mu_i,\sigma_j). \tag{3.14}$$

Parameters μ_i and σ_j making the likelihood function maximum become the estimates of the parameters μ and σ.

Numerical calculations were conducted. The programming code is shown in Appendix 3C. Numerical examples applying the above-mentioned estimation method are given below. A random number sequence $\boldsymbol{x}=x_1, x_2, \ldots, x_N$ of the length N=50 is generated from a normal distribution with the parameters of μ=0、σ=1:

$$P(x \mid \mu,\sigma) = \frac{1}{\sqrt{2\pi\sigma^2}} \exp\left(-\frac{(x-\mu)^2}{2\sigma^2}\right). \tag{3.15}$$

Fig. 3.5 shows the random sequence and a comparison of the approximate and true probability distributions.

The values of the probability calculated by equation (3.13) are shown in Table 3.2. Since the probability takes the maximum at μ=0 and σ=1, we consider the values as the estimates. The results are correct.

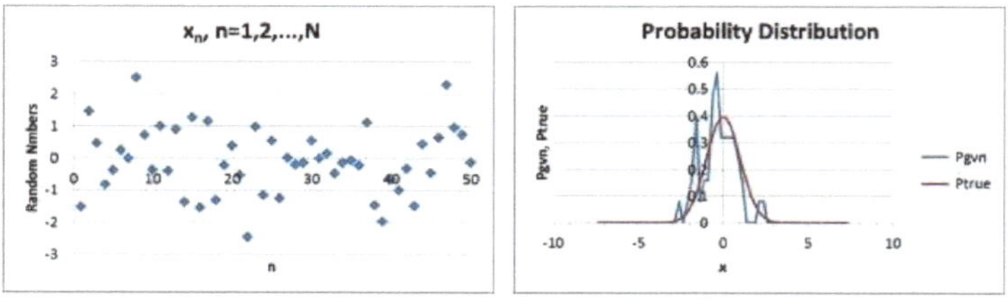

(a) Random sequence (b) Probability distribution

Fig. 3.5 A random sequence and probability distribution from normal distribution

Table 3.2 Calculation results of the reverse probability

	σ=0.875	σ=1	σ=1.125	σ=1.25	σ=1.375	σ=1.5	σ=1.625	σ=1.75
μ=-0.75	0	0.000003	0.000022	0.000035	0.000021	0.000007	0.000002	0
μ=-0.5	0.000156	0.002803	0.005469	0.00307	0.000841	0.000152	0.000022	0.000003
μ=-0.25	0.023711	0.131224	0.114232	0.035999	0.006432	0.00084	0.000093	0.00001
μ=0	0.06083	0.269952	0.201982	0.057119	0.00942	0.001157	0.000122	0.000012
μ=0.25	0.002634	0.0244	0.030235	0.012266	0.002642	0.000398	0.000049	0.000006
μ=0.5	0.000002	0.000097	0.000383	0.000356	0.000142	0.000034	0.000006	0.000001
μ=0.75	0	0	0	0.000001	0.000001	0.000001	0	0

3.3. Gamma distribution

Gamma distribution has also two positive parameters, that is, the shape parameter κ=1 and scale parameter θ =1.5 :

$$P(x\,|\,\kappa,\theta)=\frac{1}{\Gamma(\kappa)\theta^{\kappa}}x^{\kappa-1}\exp^{-x/\theta} \quad \text{for } x>0. \tag{3.16}$$

Let the candidates of the parameters be, for example

$$\begin{aligned}\kappa_i&=0.125+0.125i, \quad i=0,1,\cdots,24,\\ \theta_j&=0.125+0.125j, \quad j=0,1,\cdots,24.\end{aligned} \tag{3.17}$$

If we make similar assumptions, equation (3.13) for normal distribution becomes

$$P(\kappa_i,\theta_j\,|\,\mathbf{x}) = \frac{P(\mathbf{x}\,|\,\kappa_i,\theta_j)}{\sum_{i=1}^{I}\sum_{j=1}^{J} P(\mathbf{x}\,|\,\kappa_i,\theta_j)}. \tag{3.18}$$

The likelihood function $P(\mathbf{x}|\kappa_i,\theta_j)$ can be calculated by

$$P(\mathbf{x}\,|\,\kappa_i,\theta_j) = \prod_{n=1}^{N} P(x_n\,|\,\kappa_i,\theta_j). \tag{3.19}$$

Parameters κ_i and θ_j making equation (3.18) maximum become the estimates of the parameter κ and θ. This is nothing but the maximum likelihood method.

A numerical example of the above-mentioned estimation method is shown below. Suppose that a random sequence $\boldsymbol{x}=x_1, x_2, \ldots, x_N$ be generated by gamma distribution given by equation (3.26) with $\kappa=1$, $\theta=1.5$ and $N=200$. Fig. 3.6 shows the random sequence and a comparison of the approximate and true probability distributions.

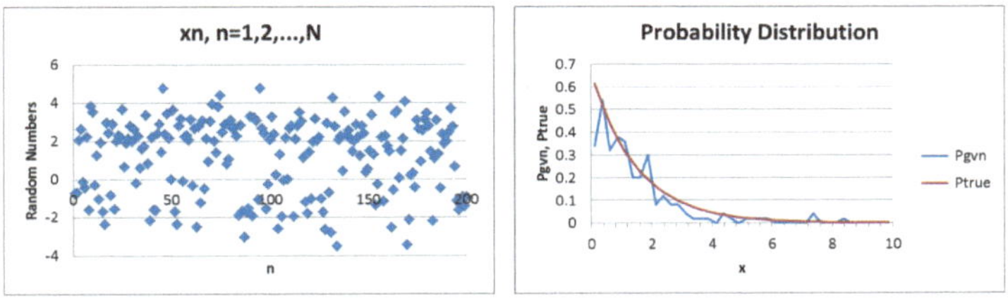

(a) Random sequence (b) Probability distribution

Fig. 3.6 A random sequence and probability distribution from gamma distribution.

The values of the probability calculated by (3.18) are shown in Table 3.3. Since the probability takes the maximum at $\kappa=1$ and $\theta=1.5$, we consider the values as the estimates. The results are correct.

Table 3.3 Calculation results of the reverse probability.

	θ=1.25	θ=1.375	θ=1.5	θ=1.625	θ=1.75	θ=1.875	θ=2.0	θ=2.125
κ=0.75	0	0	0	0.000018	0.000196	0.000749	0.001292	0.001205
κ=0.875	0.000001	0.000267	0.007728	0.04341	0.072789	0.049475	0.017012	0.003484
κ=1	0.002883	0.074423	0.245057	0.186109	0.048935	0.005927	0.000406	0.000018
κ=1.125	0.04973	0.118464	0.044303	0.004549	0.000188	0.000004	0	0
κ=1.25	0.011016	0.002422	0.000103	0.000001	0	0	0	0
κ=1.375	0.000057	0.000001	0	0	0	0	0	0
κ=1.5	0	0	0	0	0	0	0	0

3.4. How to treat when the calculated values become too small because of the product of too many probabilities

We must calculate many products of the probabilities in the Bayesian estimation. Since the probability is less than or equal to 1, the products of many probabilities make underflows. The denominator of the reverse probability does not affect the order of the size of the reverse probabilities. Hence, if we compare the logarithm of the likelihood function given by equations (3.6), (3.14) and (3.19), the maximum of the reverse function can be determined. We can prevent the underflow by taking the logarithm of the probabilities.

As an example, we consider the problem of determining the parameters of normal distribution discussed in section 3.2. We consider the logarithm of the likelihood function given by (3.14):

$$\log P(\mathbf{x} \mid \mu_i, \sigma_j) = \sum_{n=1}^{N} \log P(x_n \mid \mu_i, \sigma_j). \tag{3.20}$$

In the following calculation, a function f given by equation (3.21):

$$\begin{aligned} f(\mathbf{x} \mid \mu_i, \sigma_j) &= \exp\left[\log\left\{\prod_{n=1}^{N} \frac{P(x_n \mid \mu_i, \sigma_j)}{\max_{i',j'}\left(P(x_n \mid \mu_{i'}, \sigma_{j'})\right)}\right\}\right] \\ &= \exp\left[\sum_{n=1}^{N}\left\{\log P(x_n \mid \mu_i, \sigma_j) - \max_{i',j'}\left(\log P(x_n \mid \mu_i, \sigma_j)\right)\right\}\right] \end{aligned} \tag{3.21}$$

is used instead of equation (3.20). The calculation results for the same numerical example as in section 3.2 are shown in Table 3.4. Since μ=0 and σ=1 make f maximum, these values is considered the estimate of the parameter μ and σ.

Table 3.4 Results f given by equation (3.21)

	σ=0.75	σ=0.875	σ=1	σ=1.125	σ=1.25	σ=1.375	σ=1.5
μ=-1	0	0	0	0	0	0	0
μ=-0.75	0	0	0.00001	0.000082	0.000131	0.000078	0.000025
μ=-0.5	0.000001	0.000578	0.010382	0.02026	0.011374	0.003116	0.000563
μ=-0.25	0.000477	0.087833	0.486102	0.423157	0.133353	0.023828	0.003111
μ=0	0.00172	0.225336	1	0.748213	0.21159	0.034896	0.004287
μ=0.25	0.000024	0.009758	0.090386	0.112	0.045436	0.009787	0.001473
μ=0.5	0	0.000007	0.000359	0.001419	0.00132	0.000526	0.000126
μ=0.75	0	0	0	0.000002	0.000005	0.000005	0.000003

3.5. Compound distribution

For the solution to the problem in the present section, we need a large number of data. We face difficulties in the numerical calculations since the underflows discussed in section 3.4 occur and the calculations can't be continued. When the data number N is smaller than or equal to 200, we calculate the likelihood function using the conventional method. However, when N is bigger than 200, we take the logarithm of the likelihood function as discussed in the previous section, since the large or small relationship does not change, if we take logarithm.

We consider a compound distribution consisting of several probability distributions. As an example, we consider a compound distribution of two normal distributions. Let the parameters of the two distributions be (μ_1,σ_1) and (μ_2,σ_2), and the mixing ratio be a :

$$P(x\,|\,\mu_1,\sigma_1,\mu_2,\sigma_2,a)$$
$$=a\frac{1}{\sqrt{2\pi\sigma_1^2}}\exp\left(-\frac{(x-\mu_1)^2}{2\sigma_1^2}\right)+(1-a)\frac{1}{\sqrt{2\pi\sigma_2^2}}\exp\left(-\frac{(x-\mu_2)^2}{2\sigma_2^2}\right). \tag{3.22}$$

As the candidates of the parameters, we consider

$$\mu_{1i}=-3+\frac{6}{P}i,\quad \sigma_{1i}=0.125+\frac{3}{P}i,\quad i=0,1,\cdots,P; \tag{3.23a}$$

$$\mu_{2i}=-3+\frac{6}{P}i,\quad \sigma_{2i}=0.125+\frac{3}{P}i,\quad i=0,1,\cdots,P; \tag{3.23b}$$

$$a_i=+\frac{1}{P}i,\quad i=0,1,\cdots,P. \tag{3.23c}$$

If we assume that the prior probabilities are all equal, we then have an expression of the reverse probability similar to equation (3.13) in the case of a single normal distribution:

$$P(\mu_{1i},\sigma_{1j},\mu_{2k},\sigma_{2l},a_m\,|\,\mathbf{x})=\frac{P(\mathbf{x}\,|\,\mu_{1i},\sigma_{1j},\mu_{2k},\sigma_{2l},a_m)}{\sum_{i=1}^{I}\sum_{j=1}^{J}\sum_{k=1}^{K}\sum_{l=1}^{L}\sum_{m=1}^{M}P(\mathbf{x}\,|\,\mu_{1i},\sigma_{1j},\mu_{2k},\sigma_{2l},a_m)}. \tag{3.24}$$

The likelihood function $P(\mathbf{x}\,|\,\mu_{1i},\sigma_{1j},\mu_{2k},\sigma_{2l},a_m)$ is given by

$$P(\mathbf{x}\,|\,\mu_{1i},\sigma_{1j},\mu_{2k},\sigma_{2l},a_m)=\prod_{n=1}^{N}P(x_n\,|\,\mu_{1i},\sigma_{1j},\mu_{2k},\sigma_{2l},a_m). \tag{3.25}$$

The parameter $\mu_{1i},\sigma_{1j},\mu_{2k},\sigma_{2l},a_m$ that makes equation (3.24) maximum becomes the estimate.

First, we show a numerical result with N=200 below. We set the parameters $\mu_1=-1.2, \sigma_1=1.025, \mu_2=2.4, \sigma_2=0.875$ and $a=0.3$. A random sequence $\boldsymbol{x}=x_1, x_2, \ldots, x_N$ is generated from the compound distribution. In Fig. 3.7(a) , the random sequence is shown. In Fig. 3.7(b), the true and approximate probability distribution is shown. The approximate probability distribution means the distribution calculated from the random sequence. The maximums of the reverse probability occurred at

$$\mu_1 = -1.2, \sigma_1 = 1.025, \mu_2 = 2.4, \sigma_2 = 0.875, a = 0.35 \quad \text{and}$$
$$\mu_1 = 2.4, \sigma_1 = 0.875, \mu_2 = -1.2, \sigma_2 = 1.025, a = 0.65.$$

The parameters of each probability distribution are estimated correctly, but the estimate of the mixing ratio is not accurate. The correct estimation of the mixing ratio seems difficult with N=200.

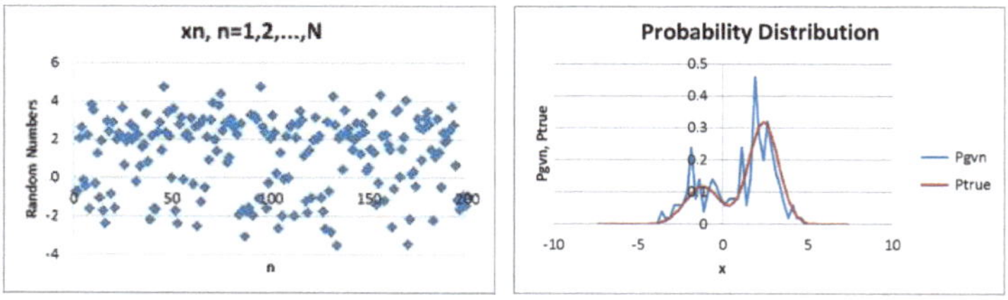

(a) Random sequence (b) Probability distribution

Fig. 3.7 A random sequence and the probability distribution.

We must conduct the numerical calculations with $N>200$ in order to estimate the mixing ratio a correctly. For this purpose, we apply the method of using a logarithm of the likelihood function instead of the likelihood function itself as discussed in section 3.4. If we take the logarithm of equation (3.25), we have

$$\log P(\mathbf{x} \mid \mu_{1i}, \sigma_{1j}, \mu_{2k}, \sigma_{2l}, a_m) = \sum_{n=1}^{N} \log P(x_n \mid \mu_{1i}, \sigma_{1j}, \mu_{2k}, \sigma_{2l}, a_m). \qquad (3.26)$$

Table 3.5 gives the results. When N=800, the correct result is given. A probability distribution obtained approximately from the frequency distribution with N=800 and the true probability distribution are also given in Fig 3.8.

Table 3.5 Estimation results using equation (3.26).

N	μ_1	σ_1	μ_2	σ_2	a	$\log P(\mathbf{x}\|\ldots)$ Biggest	2nd Biggest
200	-1.2	1.025	2.4	0.875	0.35	-385.932	-386.873
300	-1.2	1.025	2.4	0.875	0.35	-581.68	-583.072
400	-1.2	1.025	2.4	0.875	0.35	-774.149	-775.885
500	-1.2	1.025	2.4	0.875	0.35	-965.733	-967.112
600	-1.2	1.025	2.4	0.875	0.35	-1154.51	-1156.63
700	-1.2	1.025	2.4	0.875	0.35	-1329.84	-1330.53
800	-1.2	1.025	2.4	0.875	0.3	-1515.98	-1518.77

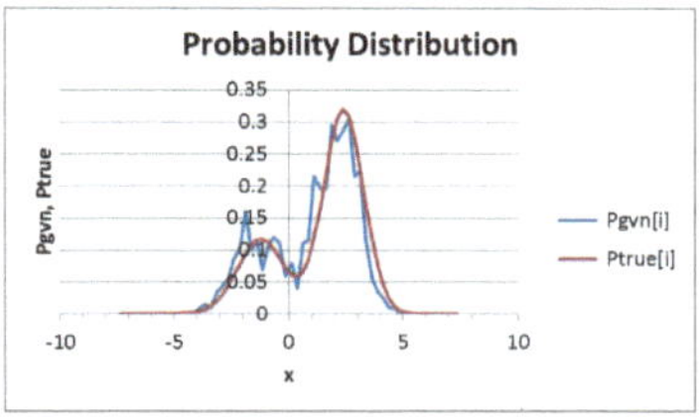

Fig. 3.8 Probability distribution (N=800).

3.6. Search of probability maximum using the mountain-climbing method

In the above discussion, we obtained the estimation by choosing the parameters making the reverse probability maximum among the candidates of the parameters set beforehand. However, we can obtain the maximum without setting the candidates beforehand.

In the following, we consider the same problem as discussed in section 3.5. However, for simplicity, we assume the parameters of the two probability distributions are given as

$$\mu_1 = -1.2, \sigma_1 = 1.025, \mu_2 = 2.8, \sigma_2 = 1.025,$$

but the mixing ratio a alone is unknown.

Furthermore, we use equation (3.26) instead of equation (3.25). If we differentiate equation (3.26), we have

$$\frac{d}{da}\log P(\mathbf{x} \mid \mu_{1i}, \sigma_{1j}, \mu_{2k}, \sigma_{2l}, a)$$

$$= \sum_{n=1}^{N} \frac{d}{da}\log P(x_n \mid \mu_{1i}, \sigma_{1j}, \mu_{2k}, \sigma_{2l}, a) = \sum_{n=1}^{N} \frac{\frac{d}{da}P(x_n \mid \mu_{1i}, \sigma_{1j}, \mu_{2k}, \sigma_{2l}, a)}{P(x_n \mid \mu_{1i}, \sigma_{1j}, \mu_{2k}, \sigma_{2l}, a)}. \quad (.3.27)$$

The calculation results are shown in Fig. 3.9. When the number of data N is increased, the mixing ratio a approaches the correct value 0.3.

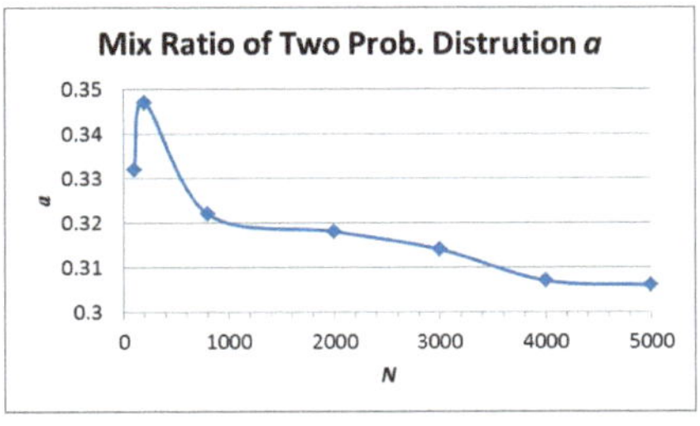

Fig. 3.9 The calculation result of the mixing ratio a using the mountain-climbing method.

In the above discussion, analytical differentiation is used. However, even if we use a numerical differentiation, we could obtain the same result. The numerical differentiation makes the calculation much easier.

In the present study, we apply the Bayesian inference to the parameter-estimation of the several probability distributions such as Bernoulli distribution, normal distribution, and gamma distribution. Furthermore, we applied the method to a compound probability distribution consisting of two normal distributions. According to the numerical calculations, satisfactory results are obtained.

If the prior probabilities are taken equal, the Bayesian inference becomes identical to the maximum likelihood method. If the data is generated from a single source as in the present case, the estimated values converge to the same values as shown in the present numerical results, as the number of the data increases. However, if the data are generated from several sources, the posterior probabilities change as the prior probabilities change.

REFERENCES 3

[1] A. Krizhevsky, I. SutskeverI, and G. Hinton. "ImageNet classification with deep convolutional neural networks," In Proc. Advances in Neural Information Processing Systems 25 1090–1098 2012.

[2] Yann LeCun1, Yoshua Bengio & Geoffrey Hinton, "Deep learning," NATURE | VOL 521 | 28 MAY 2015.

[3] M. Taki, Introduction to Deep Learning, Kodansha 2017 in Japanese.

[4] N. Matsubara, Introduction to Bayesian Statistics, Tokyo Tosho 2008 in Japanese.

[5] A. Suyama, Introduction to Mchine Learning by Bayesian Inference, Kodansha 2017 in Japanese.

[6] H. Isshiki, "Pattern Recognition by Bayesian Inference," The 63rd Joint Meeting of Automatic Control 2020.

Appendix 3A Continuus Seach of Unknown parameters

According to the Bayesian theorem, the reverse probability $P(\mu \mid \mathbf{x})$ is given by

$$P(\mu \mid \mathbf{x}) = \frac{P(\mathbf{x}, \mu)}{P(\mathbf{x})} = \frac{P(\mathbf{x} \mid \mu) P(\mu)}{P(\mathbf{x})}. \tag{A3.1}$$

Equation (A3.1) is rewritten as

$$P(\mu_i \mid \mathbf{x}) = \frac{P(\mathbf{x} \mid \mu_i)P(\mu_i)}{\int P(\mathbf{x}, \mu)d\mu} = \frac{P(\mathbf{x} \mid \mu_i)P(\mu_i)}{\int P(\mathbf{x} \mid \mu)P(\mu)d\mu}. \tag{A3.2}$$

If we assume

$$P(\mu) = \text{const.}, \tag{A3.3}$$

equation (A3.2) becomes

$$P(\mu \mid \mathbf{x}) = \frac{P(\mathbf{x} \mid \mu)}{\int P(\mathbf{x} \mid \mu)d\mu} \sim P(\mathbf{x} \mid \mu), \tag{A3.4}$$

since the denominator dose not include μ and does bot contribute in the search of the maximum. This is nothing but the likelihood method. The likelihood function $P(\mathbf{x} \mid \mu)$ could be calculated by

$$P(\mathbf{x} \mid \mu) = \prod_{n=1}^{N} P(x_n \mid \mu). \tag{A3.5}$$

The maximum of the likelihood function $P(\mathbf{x} \mid \mu)$ could be calculated by the method of the steepest ascend:

$$\mu_{new} = \mu_{old} + \frac{\partial P(\mathbf{x} \mid \mu)}{\partial \mu} d\mu. \tag{A3.6}$$

The μ that makes $P(\mu \mid \mathbf{x})$ given by equation (A3.5) the maximum gives the estimation of the parameter μ. This is nothing but the estimation by the maximum likelihood method.

Appendix 3B Code for parameter estimation of Bernoulli distribution

"Microsoft C/C++ Compiler Version 17.00.50727.1 for x86" and "Microsoft Linker Version 11.00.50727" were used for compile and link (in command window; cl source_file_name.c).

(1) Programing code: FinProbDistMyuBern.c

```
// ---------------------------------------------------------------- //
//                                                                  //
// File Name: ProbDistMyuSgmZNewX.c             2020.12.14-2020.12.16 //
// File Name: FinProbDistMyuBern.c              2020.12.18-2022.05.06 //
//                                                                  //
```

```
//    Parameter Estimation of Probability Distribution             //
//                                                                 //
// -------------------------------------------------------------- //

// ---- function --------------------------------------------------- //

#include <stdio.h>
#include <stdlib.h>
#include <string.h>
#include <math.h>

#define PI      3.14159265

void main();
void pushKey();

double factorial(int num);  // facrorial
double BinomialDist(int, int, double);  // binomial distribution
double sgn(double);          // signum function : 1 if x>0; -1 if x<0; 0 if x=0
double E(double);            // step function : 1 if x>0; 0 if x<=0

double drand();              // uniform random number in (-1,1)

double Uniform(void);                          // uniform random number
double rand_normal(double,double);             // normal radom number

int randBern(double);                         // Bernoulli random number
double Bern(int x,double);                    // Bernoulli distribution
void fsort(int);                               // sorting
void fhisto(int, double, double, double);      // histogram

double normal(double,double,double);           // normal dstribution

double max_RProb();                            // maximum value
double max_logRProb();                         // maximum of log value

// ---- variable ------------------------------------------------- //

char title_memo[5000];
```

```
int P;                          // number of pattern
int N;                          // number of Data

int N1;                         // number of sampling

//double PatData[6][6][5][6];   // Data of pattern
double Prule[10001][1001];      // Probaility of rule
double Pgvn[10001];             // given distribution
double Ptrue[10001];            // true distribution

//int Ptn[6][6];                // Input pattern

double RProb[101];              // reverse probability
double logRProb[101];           // log of reverse probability
double RProb1[101];             // exp(logRprob[]-max_logRProb[])

double AMAT[1001][2001];        // matrix

FILE *fp_inp;                   // input file pointer
FILE *fp_out;                   // output file pointer

char InputDataFile[80];         // input file name
char OutputDataFile[80];        // output file name

char buf[5000];

double x[10001];                 // x
   int M;                          // number of data
   double Value[10001];            // value
   double histo[10001];            // frequency
   double x_max;
   double x_min;
   double Dx;
   double width;

   double myu[101];
   double sgm[101];
   double myu1;
```

```
double sgm1;

int ipnt[10001];

int p_max;
int pp_max;

// ------------------------------------------------------------ //

void main()
{
    int p, pp, m, n;
    int i, j;
    double sum, tmp;

    //// open input file
    sprintf(InputDataFile, "FinProbDistMyuBern_inp.dat");

    if ((fp_inp = fopen(InputDataFile, "r")) == NULL) {
        printf("Failed in Reading Input Data File! ... %s¥n", InputDataFile);
        exit(1);
    }

  //// open output file
  sprintf(OutputDataFile, "FinProbDistMyuBern_out.csv");

  if ((fp_out = fopen(OutputDataFile, "w")) == NULL) {
      printf("Failed in Reading Output Data File! ... %s¥n", OutputDataFile);
      exit(1);
  }

  //// input from file
  fscanf(fp_inp, "%s", title_memo);

  fscanf(fp_inp, "%s %d", buf, &P);
  fscanf(fp_inp, "%s %d", buf, &N1);
  fscanf(fp_inp, "%s %lf", buf, &myu1);
```

```
printf("memo: %s¥n", title_memo);
printf("¥n");

printf("P     = %d¥n", P);
printf("N1    = %d¥n", N1);
printf("myu1  = %12.6f¥n", myu1);
printf("¥n");


fprintf(fp_out, "memo: %s¥n", title_memo);
fprintf(fp_out, "¥n");

fprintf(fp_out, "P =, %d¥n", P);
fprintf(fp_out, "N1 =, %d¥n", N1);
fprintf(fp_out, "myu1 =, %12.6f¥n", myu1);

fprintf(fp_out, "¥n");

width = x_max-x_min;

for (p = 0; p <= P; p++)
    myu[p] = 0.0+(p+0.0)*1.0/(P+0.0);

fprintf(fp_out, "p, myu[p]¥n");
for (p = 0; p <= P; p++)
    fprintf(fp_out, "%d, %12.6f¥n", p, myu[p]);
fprintf(fp_out, "¥n");
// input data

for (n = 1; n <= N1; n++)
    Value[n] = randBern(myu1)+0.0;

fprintf(fp_out, "n, Value[n]¥n");
for(n = 1; n <= N1; n++)
    fprintf(fp_out, "%d, %12.6f¥n", n, Value[n]);
fprintf(fp_out, "¥n");

fprintf(fp_out, "p, myu, RProb¥n");
for (p = 0; p <= P; p++) {
```

```
            RProb[p] = 1.0;
            for (n = 1; n <= N1; n++) {
                RProb[p] *= Bern((int)Value[n], myu[p]);
            }
            fprintf(fp_out, "%d, %lf, %lg¥n", p, myu[p], RProb[p]);
    }
    fprintf(fp_out, "¥n");

//      fprintf(fp_out, "p, myu, RProb, logRProb¥n");
    for (p = 0; p <= P; p++) {
        RProb[p] = 1.0;
        logRProb[p] = 0.0;
        for (n = 1; n <= N1; n++) {
            tmp = Bern((int)Value[n], myu[p]);
            RProb[p] *= tmp;
            logRProb[p] += log(tmp);
        }
//          fprintf(fp_out, "%d, %lf, %lg, %lg¥n", p, myu[p], RProb[p], logRProb[p]);
    }
    fprintf(fp_out, "¥n");

    tmp = max_logRProb();

    fprintf(fp_out, "p, myu, RProb1¥n");
    for (p = 0; p <= P; p++) {
        RProb1[p] = exp(logRProb[p]-tmp);
        fprintf(fp_out, "%d, %lf, %lg¥n", p, myu[p], RProb1[p]);
    }
    fprintf(fp_out, "¥n");

    sum = 0.0;
    for (p = 0; p <= P; p++)
        sum += RProb1[p];

    fprintf(fp_out, "p, myu, RProbNmlzd¥n");
    for (p = 0; p <= P; p++)
        fprintf(fp_out, "%d, %lf, %lf¥n", p, myu[p], RProb1[p]/sum);
    fprintf(fp_out, "¥n");
```

```
 tmp = max_RProb();
 fprintf(fp_out, "p_max =, %d, or, myu_max =, %lf, RProb_max = %lg¥n",
         p_max, myu[p_max], tmp);

 fclose(fp_out);

    pushKey();
}

// ------------------------------------------------------------ //

void pushKey()
{
    printf("¥n      Push Return Key! ");
    getchar();
    getchar();
}

// ------------------------------------------------------------ //

double factorial(int num)
{
    int i;
    double fact;

    // factorial //
    fact = 1.0;

    for(i=1; i<=num; ++i)
        fact = fact * (i+0.0);

    return fact;
}

// ------------------------------------------------------------ //

// binomial distribution
double BinomialDist(int m, int M, double myu)
{
```

```
    return factorial(M)/factorial(m)/factorial(M-m)*pow(myu,m+0.0)*pow(1.0-myu,M-m+0.0);
}

// ---------------------------------------------------------------- //

double sgn(double x)
{
    if (x > 0)
        return 1.0;
    else if (x < 0)
        return -1.0;
    else
        return 0.0;
}

// ---------------------------------------------------------------- //

double E(double x)
{
    if (x > 0)
        return x;
    else
        return 0.0;
}

// ---------------------------------------------------------------- //

double drand()
{
    return 2.0*(((double)rand())/((double)RAND_MAX)-0.5);
}

// ---------------------------------------------------------------- //

double rand_normal( double myu, double sigma )
{
//      double z=sqrt( -2.0*log(Uniform()) ) * sin( 2.0*M_PI*Uniform() );
  double z=sqrt( -2.0*log(Uniform()) ) * sin( 2.0*PI*Uniform() );
  return myu + sigma*z;
```

```
}

// ------------------------------------------------------------ //

double Uniform( void )
{
 static int x=10;
 int a=1103515245,b=12345,c=2147483647;
 x = (a*x + b)&c;

 return ((double)x+1.0) / ((double)c+2.0);
}

// ------------------------------------------------------------ //

int randBern(double myu)
{
    if (Uniform() <= myu)
 return 1;
    else
        return 0;
}

// ------------------------------------------------------------ //

double Bern(int x,double myu)
{
    if (x == 1)
        return myu;
    else
        return 1.0-myu;
}

// ------------------------------------------------------------ //

void fsort(int n )
{
 long   i, j;
 for(i=0;i<n-1;i++){
```

```
        for(j=i+1;j<n;j++){
                if(Value[j] > Value[i]){
                        double temp = Value[i];
                        Value[i] = Value[j];
                        Value[j] = temp;
                }
        }
 }
}

// ---------------------------------------------------------------- //

void fhisto(int n, double x_min, double x_max, double Dx)
{
 int i, j, k;
 double u_limit, l_limit;

 fprintf(fp_out, "x_max =, %12.6f,     x_min =, %12.6f,     width/Dx =, %12.6f¥n¥n", x_max, x_min,
width/Dx);

 for(i=0;i<(int)(width/Dx);i++){
        histo[i] = 0;
 }

 u_limit = (x_min + Dx);
 l_limit =  x_min;

 for(k=0;k<(int)(width/Dx)+0.000001;k++){
        for(i=0;i<n;i++){
                if(Value[i] < u_limit && Value[i] >= l_limit){
                        histo[k] += 1.0;
        }
        }
        u_limit = u_limit+Dx;
        l_limit = l_limit+Dx;
 }

//      for(j=0;j<(int)width/Dx+0.000001;j++){
//              fprintf(fp_out, "%12.6f <= Value < %12.6f     histo[%d] = %12.6f¥n",
```

```
//                                  (x_min+j*Dx), (x_min+(j+1)*Dx), j, histo[j]);
//       }
//         fprintf(fp_out, "¥n");
}

// ------------------------------------------------------------ //

double normal(double x, double myu, double sgm)              // normal distribution
{
    return 1.0/sqrt(2.0*PI*sgm*sgm)*exp(-(x-myu)*(x-myu)/2.0/sgm/sgm);
}

// ------------------------------------------------------------ //

double max_RProb()                  // maximum
{
    int p;
    double tmp;

    tmp = -1.0E-200;
    for (p = 0; p <= P; p++)
        if (tmp <= RProb[p]) {
            tmp = RProb[p];
            p_max = p;
        }

    return tmp;
}

// ------------------------------------------------------------ //

double max_logRProb()          .       // log maximum
{
    int p, pp;
    double tmp;

    tmp = -1.0E-200;
    for (p = 0; p <= P; p++)
        if (tmp <= logRProb[p]) {
```

```
            tmp = logRProb[p];
            p_max = p;
        }
    return tmp;
}

// ------------------------------------------------------------------ //
```

(2) **Input file: FinProbDistMyuBern_inp.dat**

```
ProbDistParam_20201220

P          20
N1         50
myu1       0.35
```

Appendix 3C Code for parameter estimation of Normal distribution

"Microsoft C/C++ Compiler Version 17.00.50727.1 for x86" and "Microsoft Linker Version 11.00.50727" were used for compile and link (in command window; cl source_file_name.c).

(1) Programing code: ProbDistMyuSgmZNewX.c

```
// ------------------------------------------------------------------ //
//                                                                    //
// File Name: BayesPattern.c                 2020.08.10-2020.08.17 //
// File Name: BayesPatternX.c                2020.08.17-2020.08.18 //
// File Name: BayesPatternY.c                2020.08.18-2020.08.18 //
// File Name: BayesPatternZ.c                2020.08.18-2020.08.18 //
// File Name: ProbDistMyuZ.c                 2020.10.25-2020.10.27 //
// File Name: ProbDistMyuZNew.c              2020.12.11-2020.12.13 //
// File Name: ProbDistMyuZNewX.c             2020.12.13-2020.12.14 //
// File Name: ProbDistSgmZNewX.c             2020.12.14-2020.12.14 //
// File Name: ProbDistMyuSgmZNewX.c          2020.12.14-2022.05.06 //
//                                                                    //
//    Parameter Estimation of Probability Distribution                //
//                                                                    //
// ------------------------------------------------------------------ //
```

```
// ----- function ---------------------------------------------------- //

#include <stdio.h>
#include <stdlib.h>
#include <string.h>
#include <math.h>

#define PI      3.14159265

void main();
void pushKey();

double sgn(double);          // signum function : 1 if x>0; -1 if x<0; 0 if x=0
double E(double);            // step function : 1 if x>0; 0 if x<=0

double drand();              // uniform random number in (-1,1)

double Uniform( void );                       // uniform random number
double rand_normal(double,double);            // normal random number

void fsort(int);                              // sorting
void fhisto(int, double, double, double);     // histogram

double normal(double,double,double);          // normal distribution

double max_RProb();                           // maximum value
double max_logRProb();                        // maximum of log value

// ----- variable -------------------------------------------------- //

char title_memo[5000];

int P;                            // number of pattern
int N;                            // number of Data

int N1;                           // number of sampling

//double PatData[6][6][5][6];     // Data of pattern
double Prule[10001][1001];        // Probaility of rule
```

```
double Pgvn[10001];                // given distribution
double Ptrue[10001];               // true distribution

//int Ptn[6][6];                   // Input pattern

double RProb[101][101];            // reverse probability
double logRProb[101][101];         // log of reverse probability
double RProb1[101][101];           // exp(logRprob[]-max_logRProb[])

double AMAT[1001][2001];           // matrix

FILE *fp_inp;                      // input file pointer
FILE *fp_out;                      // output file pointer

char InputDataFile[80];            // input file name
char OutputDataFile[80];           // outputfile name

char buf[5000];

double x[10001];                    // x

int Num;                           // number of data
double Value[10001];               // value
double histo[10001];               // frequency
double x_max;
double x_min;
double Dx;
double width;

double myu[101];
double sgm[101];
double myu1;
double sgm1;

int ipnt[10001];

int p_max;
int pp_max;
```

```
// ------------------------------------------------------------ //

void main()
{
    int p, pp, n;
    int i, j;
    double sum, tmp;

    //// open input file
    sprintf(InputDataFile, "ProbDistMyuSgmZNewX_inp.dat");

    if ((fp_inp = fopen(InputDataFile, "r")) == NULL) {
        printf("Failed in Reading Input Data File! ... %s¥n", InputDataFile);
        exit(1);
    }

    //// open output file
    sprintf(OutputDataFile, "ProbDistMyuSgmZNewX_out.csv");

    if ((fp_out = fopen(OutputDataFile, "w")) == NULL) {
        printf("Failed in Reading Output Data File! ... %s¥n", OutputDataFile);
        exit(1);
    }


    //// input from file
    fscanf(fp_inp, "%s", title_memo);

    fscanf(fp_inp, "%s %d", buf, &Num);
    fscanf(fp_inp, "%s %lf", buf, &x_min);
    fscanf(fp_inp, "%s %lf", buf, &x_max);
    fscanf(fp_inp, "%s %lf", buf, &Dx);

    fscanf(fp_inp, "%s %d", buf, &P);
    fscanf(fp_inp, "%s %d", buf, &N);
    fscanf(fp_inp, "%s %d", buf, &N1);
    fscanf(fp_inp, "%s %lf", buf, &myu1);
    fscanf(fp_inp, "%s %lf", buf, &sgm1);
```

```
printf("memo: %s¥n", title_memo);
printf("¥n");

printf("Num   = %d¥n", Num);
printf("x_min = %12.6f¥n", x_min);
printf("x_max = %12.6f¥n", x_max);
printf("Dx    = %12.6f¥n", Dx);
printf("¥n");

printf("P     = %d¥n", P);
printf("N     = %d¥n", N);
printf("N1    = %d¥n", N1);
printf("myu1  = %12.6f¥n", myu1);
printf("sgm1  = %12.6f¥n", sgm1);
printf("¥n");


fprintf(fp_out, "memo: %s¥n", title_memo);
fprintf(fp_out, "¥n");

fprintf(fp_out, "Num =, %d¥n", Num);
fprintf(fp_out, "x_min =, %12.6f¥n", x_min);
fprintf(fp_out, "x_max =, %12.6f¥n", x_max);
fprintf(fp_out, "Dx =, %12.6f¥n", Dx);
fprintf(fp_out, "¥n");

fprintf(fp_out, "P =, %d¥n", P);
fprintf(fp_out, "N =, %d¥n", N);
fprintf(fp_out, "N1 =, %d¥n", N1);
fprintf(fp_out, "myu1 =, %12.6f¥n", myu1);
fprintf(fp_out, "sgm1 =, %12.6f¥n", sgm1);
fprintf(fp_out, "¥n");

width = x_max-x_min;

for (p = 0; p <= P; p++) {
    myu[p] = -3.0+(p+0.0)*6.0/(P+0.0);
    sgm[p] = 0.125+(p+0.0)*3.0/(P+0.0);
```

```
    }

    fprintf(fp_out, "p, myu[p]¥n");
    for (p = 0; p <= P; p++)
        fprintf(fp_out, "%d, %12.6f¥n", p, myu[p]);
    fprintf(fp_out, "¥n");

    fprintf(fp_out, "p, sgm[p]¥n");
    for (p = 0; p <= P; p++)
        fprintf(fp_out, "%d, %12.6f¥n", p, sgm[p]);
    fprintf(fp_out, "¥n");


    // input data

    for (n = 1; n <= N1; n++)
        Value[n] = rand_normal(myu1,sgm1);

    fprintf(fp_out, "n, Value[n]¥n");
    for(n = 1; n <= N1; n++)
        fprintf(fp_out, "%d, %12.6f¥n", n, Value[n]);
    fprintf(fp_out, "¥n");

    fhisto(N1, x_min, x_max, Dx);

    for(i = 1; i <= (x_max-x_min)/Dx; i++)
        Pgvn[i] = (histo[i]+0.0);

    for(i = 1; i <= (x_max-x_min)/Dx; i++) {
        Pgvn[i] /= ((N1+0.0)*Dx);
        Ptrue[i] = normal(x_min+(i-0.5)*Dx, myu1, sgm1);
    }

    fprintf(fp_out, "i, x, Pgvn[i], Ptrue[i]¥n");
    for(i = 1; i <= (x_max-x_min)/Dx; i++)
        fprintf(fp_out, "%d, %12.6f, %12.6f, %12.6f¥n", i, x_min+(i-0.5)*Dx, Pgvn[i],
Ptrue[i]);
    fprintf(fp_out, "¥n");
```

```
fprintf(fp_out, "p, pp, myu, sgm, RProb¥n");
for (p = 0; p <= P; p++) {
    for (pp = 0; pp <= P; pp++) {
        RProb[p][pp] = 1.0;
        for (n = 1; n <= N1; n++) {
            RProb[p][pp] *= normal(Value[n], myu[p], sgm[pp]);
        }
        fprintf(fp_out, "%d, %d, %lf, %lf, %lg¥n", p, pp, myu[p], sgm[pp], RProb[p][pp]);
    }
}
fprintf(fp_out, "¥n");

fprintf(fp_out, "p: downward, pp: rightward, RProb¥n");
fprintf(fp_out, """, ");
for (pp = 0; pp <= P; pp++)
    fprintf(fp_out, "pp = %d, ", pp);
fprintf(fp_out, "¥n");
for (p = 0; p <= P; p++) {
    fprintf(fp_out, "p = %d, ", p);
    for (pp = 0; pp <= P; pp++)
        fprintf(fp_out, "%lg, ", RProb[p][pp]);
    fprintf(fp_out, "¥n");
}
fprintf(fp_out, "¥n");

sum = 0.0;
for (p = 0; p <= P; p++)
    for (pp = 0; pp <= P; pp++)
        sum += RProb[p][pp];

fprintf(fp_out, "p: downward, pp: rightward, RProbNmlzd¥n");
fprintf(fp_out, """, ");
for (pp = 0; pp <= P; pp++)
    fprintf(fp_out, "pp = %d, ", pp);
fprintf(fp_out, "¥n");
for (p = 0; p <= P; p++) {
    fprintf(fp_out, "p = %d, ", p);
    for (pp = 0; pp <= P; pp++) {
        RProb[p][pp]/= sum;
```

```
            fprintf(fp_out, "%lf, ", RProb[p][pp]);
        }
        fprintf(fp_out, "¥n");
    }
    fprintf(fp_out, "¥n");

    tmp = max_RProb();
    fprintf(fp_out, "p_max =, %d, or, myu_max =, %lf, pp_max =, %d, or, sgm_max =, %lf, RProb_max
= %lf¥n",
            p_max, myu[p_max], pp_max, sgm[pp_max], tmp);

    fclose(fp_out);

    pushKey();
}

// ------------------------------------------------------------------ //

void pushKey()
{
    printf("¥n        Push Return Key! ");
    getchar();
    getchar();
}

// ------------------------------------------------------------------ //

double sgn(double x)
{
    if (x > 0)
        return 1.0;
    else if (x < 0)
        return -1.0;
    else
        return 0.0;
}

// ------------------------------------------------------------------ //
```

```
double E(double x)
{
    if (x > 0)
        return x;
    else
        return 0.0;
}

// ------------------------------------------------------------ //

double drand()
{
    return 2.0*(((double)rand())/((double)RAND_MAX)-0.5);
}

// ------------------------------------------------------------ //

double rand_normal( double myu, double sigma )
{
//      double z=sqrt( -2.0*log(Uniform()) ) * sin( 2.0*M_PI*Uniform() );
 double z=sqrt( -2.0*log(Uniform()) ) * sin( 2.0*PI*Uniform() );
 return myu + sigma*z;
 }

// ------------------------------------------------------------ //

double Uniform( void )
{
 static int x=10;
 int a=1103515245,b=12345,c=2147483647;
 x = (a*x + b)&c;

 return ((double)x+1.0) / ((double)c+2.0);
}

// ------------------------------------------------------------ //

void fsort(int n )
{
```

```
  long    i, j;
  for(i=0;i<n-1;i++){
          for(j=i+1;j<n;j++){
                  if(Value[j] > Value[i]){
                          double temp = Value[i];
                          Value[i] = Value[j];
                          Value[j] = temp;
                  }
          }
  }
}

// ---------------------------------------------------------------- //

void fhisto(int n, double x_min, double x_max, double Dx)
{
  int i, j, k;
  double u_limit, l_limit;

  fprintf(fp_out, "x_max =, %12.6f,    x_min =, %12.6f,    width/Dx =, %12.6f¥n¥n", x_max, x_min,
width/Dx);

  for(i=0;i<(int)(width/Dx);i++){
          histo[i] = 0;
  }

  u_limit = (x_min + Dx);
  l_limit =  x_min;

  for(k=0;k<(int)(width/Dx)+0.000001;k++){
          for(i=0;i<n;i++){
                  if(Value[i] < u_limit && Value[i] >= l_limit){
                          histo[k] += 1.0;
          }
          }
          u_limit = u_limit+Dx;
          l_limit = l_limit+Dx;
  }
}
```

```
// -------------------------------------------------------------- //

double normal(double x, double myu, double sgm)             // normal distribution
{
    return 1.0/sqrt(2.0*PI*sgm*sgm)*exp(-(x-myu)*(x-myu)/2.0/sgm/sgm);
}

// -------------------------------------------------------------- //

double max_RProb()                  // maximum value
{
    int p, pp;
    double tmp;

    tmp = -1.0E-200;
    for (p = 0; p <= P; p++)
        for (pp = 0; pp <= P; pp++)
            if (tmp <= RProb[p][pp]) {
                tmp = RProb[p][pp];
            p_max = p;
            pp_max = pp;
        }

    return tmp;
}

// -------------------------------------------------------------- //

double max_logRProb()                  // maximum of log value
{
    int p, pp;
    double tmp;

    tmp = -1.0E-200;
    for (p = 0; p <= P; p++)
        for (pp = 0; pp <= P; pp++)
            if (tmp <= logRProb[p][pp]) {
                tmp = logRProb[p][pp];
```

```
            p_max = p;
            pp_max = pp;
        }

    return tmp;
}

// ------------------------------------------------------------------ //
```

(2) Input file: ProbDistMyuSgmZNewX_inp.dat

```
ProbDistParam_20201214

Num         100
x_min       -7.5
x_max       +7.5
Dx          0.25

P           24
N           20
N1          50
myu1        0.0
sgm1        1.0
```

4. INTERPOLATION OF FUNCTION AND IDENTIFICATION OF DIFFERENTIAL EQUATION

In this chapter, we discuss the application of the Bayesian inference to the problems of functional interpolation and differential equation identification. These problems can be solved in the traditional way, but I think the Bayesian inference provides a new perspective on these problems. It also gives some suggestions regarding information processing in the human brain.

In this chapter as well, the solution must be found by searching for the maximum value. There are various possible methods for searching for the maximum value, but the simplest and easiest to understand is to list multiple discrete value candidates and select from them. This chapter also relies mainly on this method with an emphasis on comprehensibility. For continuous variables, the Steepest Ascent method is used generally. Since the probability takes a value between 0 and 1, the value of the likelihood function becomes very small and may cause underflow. In such cases, it is effective to take the logarithm. Even if the logarithm is taken, the magnitude relationship does not change. Section (C) below describes taking the logarithm and the steepest ascent method. At that time, the partial differential value by the unknown variable of the objective function is required. It is convenient to use the numerical derivative instead of the analytical derivative.

4.1. Function interpolation by the Bayesian inference

4.1.1. Linear interpolation

(A) Theory

As shown in Fig. 4.1, N data points as

$$(x_n, y_n),\ n = 1, 2, \cdots, N \tag{4.1}$$

are given. We interpolate N data points with a linear equation $y = ax + b$. We consider the difference ε_n between the data at point n and the value of the interpolation function. Namely,

$$y_n = ax_n + b + \varepsilon_n. \tag{4.2}$$

With respect to ε_n, we consider a random number following a normal distribution with a mean of 0 and with a variance of σ^2. That is

$$\varepsilon_n = y_n - (ax_n + b) \sim \frac{1}{\sqrt{2\pi\sigma^2}} \exp\left(-\frac{\varepsilon_n^{\ 2}}{2\sigma^2} \right). \tag{4.3}$$

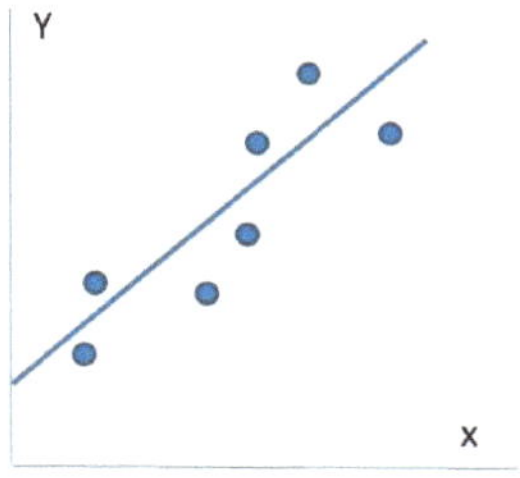

Fig. 4.1 Interpolation by a linear function.

At this time, let $P(\varepsilon_n \mid a,b,\sigma)$ be the probability of the value that ε_n takes when a, b, σ is given. We have

$$P(\varepsilon_n \mid a,b,\sigma) = \frac{1}{\sqrt{2\pi\sigma^2}} \exp\left(-\frac{\varepsilon_n^{\ 2}}{2\sigma^2} \right) = \frac{1}{\sqrt{2\pi\sigma^2}} \exp\left(-\frac{(y_n - ax_n - b)^2}{2\sigma^2} \right). \quad (4.4)$$

Because the difference ε_n is considered independent of each other, we obtain

$$P(\varepsilon_1, \varepsilon_2, \cdots, \varepsilon_N \mid a,b,\sigma) = \prod_{n=1}^{N} P(\varepsilon_n \mid a,b,\sigma). \quad (4.5)$$

Applying Bayes' theorem, the reverse probability becomes

$$\begin{aligned} P(a,b,\sigma \mid \varepsilon_1, \varepsilon_2, \cdots, \varepsilon_N) &= \frac{P(\varepsilon_1, \varepsilon_2, \cdots, \varepsilon_N, a, b, \sigma)}{P(\varepsilon_1, \varepsilon_2, \cdots, \varepsilon_N)} \\ &= \frac{P(\varepsilon_1, \varepsilon_2, \cdots, \varepsilon_N, a, b, \sigma)}{\int da \int db \int d\sigma P(\varepsilon_1, \varepsilon_2, \cdots, \varepsilon_N, a, b, \sigma)} \\ &= \frac{P(\varepsilon_1, \varepsilon_2, \cdots, \varepsilon_N \mid a,b,\sigma) P(a) P(b) P(\sigma)}{\int da \int db \int d\sigma P(\varepsilon_1, \varepsilon_2, \cdots, \varepsilon_N \mid a,b,\sigma) P(a) P(b) P(\sigma)}. \end{aligned} \quad (4.6)$$

When considering a, b, σ as a discrete value instead of a continuous value in the numerical calculation, equation (4.6) becomes

$$\begin{aligned} &P(a_i, b_j, \sigma_k \mid \varepsilon_1, \varepsilon_2, \cdots, \varepsilon_N) \\ &= \frac{P(\varepsilon_1, \varepsilon_2, \cdots, \varepsilon_N \mid a_i, b_j, \sigma_k) P(a_i) P(b_j) P(\sigma_k)}{\sum_{i}^{I} \sum_{j}^{J} \sum_{k}^{K} P(\varepsilon_1, \varepsilon_2, \cdots, \varepsilon_N \mid a_i, b_j, \sigma_k) P(a_i) P(b_j) P(\sigma_k)}. \end{aligned} \quad (4.7)$$

$P(a_i)$, $P(b_j)$, and $P(\sigma_k)$ are called the prior probability. $P(\varepsilon_1, \varepsilon_2, \cdots, \varepsilon_N \mid a_i, b_j, \sigma_k)$ and $P(a_i, b_j, \sigma_k \mid \varepsilon_1, \varepsilon_2, \cdots, \varepsilon_N)$ are the likelihood function, and the posterior probability, respectively.

If all prior probabilities are equal, respectively, that is

$$P(a_i) = const, \text{ for } \forall i;\ P(b_j) = const, \text{ for } \forall j;\ P(\sigma_k) = const \text{ for } \forall k, \qquad (4.8)$$

equation (4.7) becomes

$$P(a_i, b_j, \sigma_k \mid \varepsilon_1, \varepsilon_2, \cdots, \varepsilon_N) = \frac{P(\varepsilon_1, \varepsilon_2, \cdots, \varepsilon_N \mid a_i, b_j, \sigma_k)}{\sum_i^I \sum_j^J \sum_k^K P(\varepsilon_1, \varepsilon_2, \cdots, \varepsilon_N \mid a_i, b_j, \sigma_k)}. \qquad (4.9)$$

This is nothing but the likelihood method. For the problem of linear interpolation, the assumption in equation (4.8) seems appropriate. Therefore, in the following numerical example, the estimation of a_i, b_j, σ_k is performed using equation (4.9).

(B) Numerical example 1: Discrete approach

In order to verify the above theory, numerical calculations were performed and comparisons were made with conventional methods. First, create the equivalent of experimental data or virtual data. Therefore, a random number sequence according to equations (4.2) and (4.3) is generated. In the example below, we used

$$0 < x < 1,\ a = b = 1,\ \sigma = 0.2,\ N = 25, \qquad (4.10a)$$

$$0 < x < 1,\ a = b = 1,\ \sigma = 0.4,\ N = 50. \qquad (4.10b)$$

The following are used as candidates for unknown discrete parameters a_i, b_j, σ_k:

$$\begin{aligned} a_i &= -2 + i \times 0.2,\ i = 0, 1, \cdots, 20, \\ b_j &= -0.5 + j \times 0.1,\ j = 0, 1, \cdots, 20, \\ \sigma_k &= 0.05 + k \times 0.05,\ k = 0, 1, \cdots, 20. \end{aligned} \qquad (4.11)$$

(B-1) When the coordinate sequence $x_n,\ n = 1, 2, \cdots, N$ is taken regularly

In the following, Fig. 4.2 shows the result when the coordinate sequence $x_n,\ n = 1, 2, \cdots, N$ is taken regularly. The three figures on the left side are created according to equation (4.10a), and the three figures on the right side are created according to equation (4.10b). The two figures on the left and right at the top show a sequence of data points. The two figures on the left and right in the middle show the search results for the maximum probability. In this example, we have

$$i = 14,\ j = 16,\ k = 4 \text{ or } a_{Bayes} = 0.8,\ b_{Bayes} = 1.1,\ \sigma_{Bayes} = 0.25, \qquad (4.12a)$$

$$i = 15,\ j = 15,\ k = 7 \text{ or } a_{Bayes} = 1.0,\ b_{Bayes} = 1.0,\ \sigma_{Bayes} = 0.40, \qquad (4.12b)$$

where the suffix "Bayes" means the maximum of the reverse probability, that is, the Bayesian inference. The two figures on the left and right at the bottom show the data point sequence $y_n = ax_n + b + \varepsilon_n$, the base linear function $y_{Base} = ax + b$ used when the data point sequence was generated, the interpolation result by Bayesian inference $y_{Bayes} = a_{Bayes}x + b_{Bayes}$, and the interpolation result $y_{LSM} = a_{LSM}x + b_{LSM}$ by the least-squares method. It can be seen that there is not much difference between y_{Base}, y_{Bayes} and y_{LSM}.

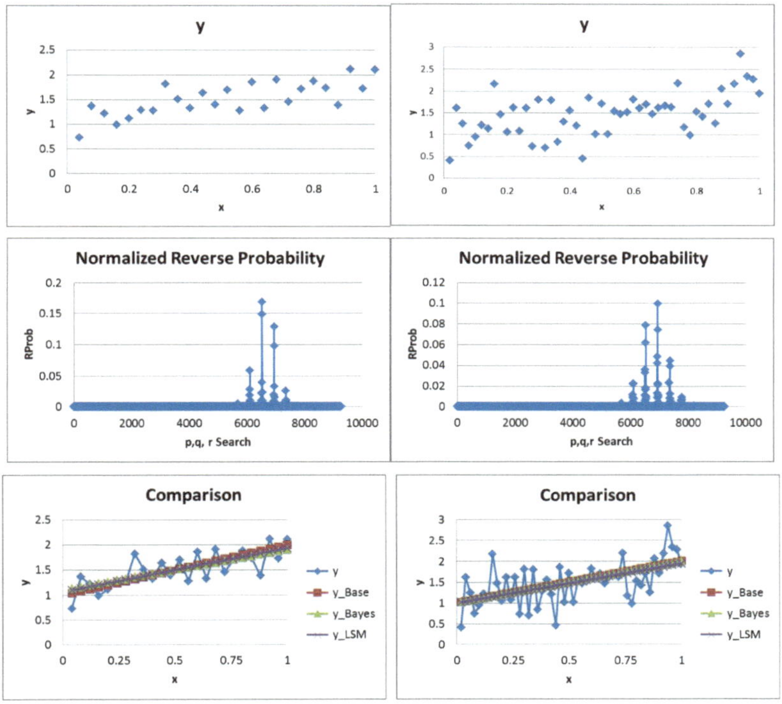

(a) N=25 (b) N=50

Fig. 4.2 A numerical example of interpolation by a linear function (x-direction is a regular point sequence).

In the above estimation, even if the same probability distribution is assumed for the error ε_n in equation (4.7), the value of ε_n is changed every time data is generated, so the parameter estimation result also changes the value every time. A large number of trials can be used to estimate its statistical properties.

Fig. 4.3 shows the statistical properties of the estimated parameters for the cases where $N = 25$, $N = 50$, and $N = 100$, where M is the number of trials. It can be seen that as N increases, the variation in parameters decreases. A regular sequence of points is used in the x-direction. Similar results are obtained by using an irregular sequence of points in the x-direction. The average and standard deviation of the parameters estimated from the results in Fig. 4.3 can be obtained.

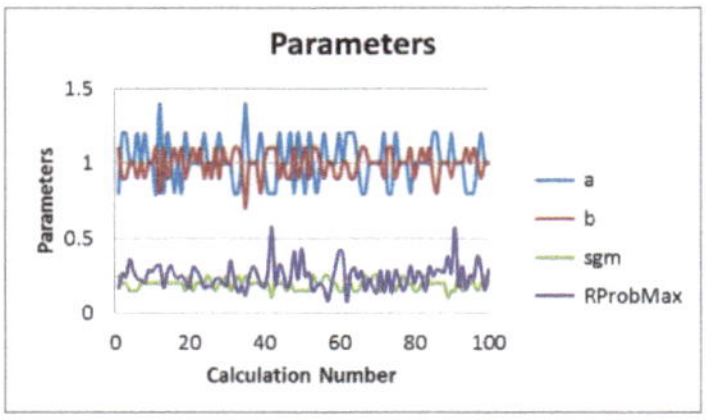

N=25, M=100

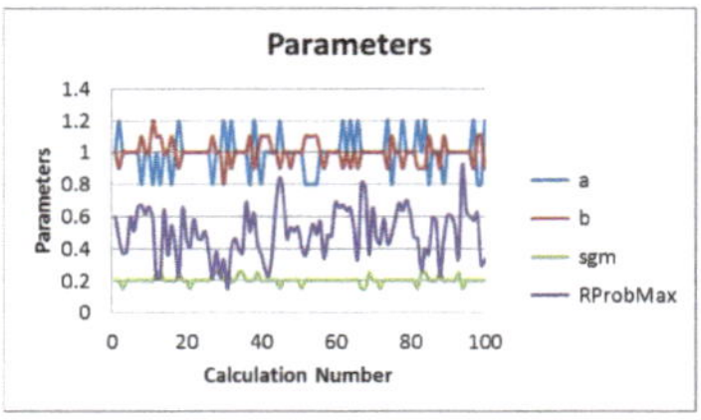

N=50, M=100

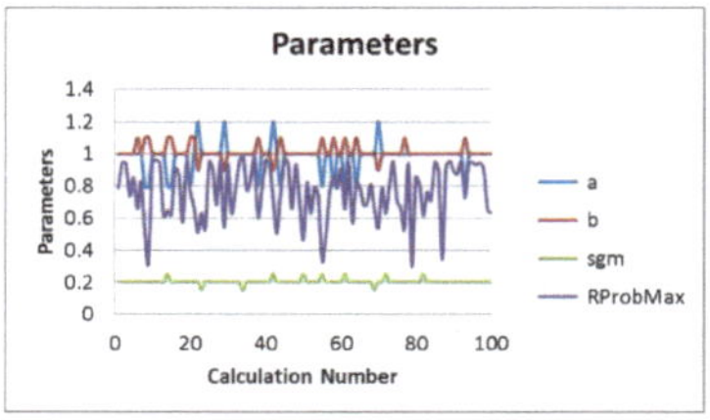

N=100, M=100

Fig. 4.3 Statistical properties of the estimated parameters.

(B-2) When the coordinate sequence $x_n,\ n=1,2,\cdots,N$ is taken irregularly

In the following, Fig. 4.4 shows the case where the coordinate sequence is taken irregularly. The three figures on the left follow equation (4.10a), and the three figures on the right follow equation (4.10b) to create a sequence of data points. The two figures on the left and right at the top show a sequence of data points. The two figures on the left and right in the middle show the search results for the maximum probability. In this example, we obtained

$$i=14,\ j=16,\ k=4 \text{ or } a_{Bayes}=0.8, b_{Bayes}=1.1, \sigma_{Bayes}=0.25, \tag{4.13a}$$

$$i=14,\ j=16,\ k=7 \text{ or } a_{Bayes}=0.8, b_{Bayes}=1.1, \sigma_{Bayes}=0.40. \tag{4.13b}$$

The suffix "Bayes" mean the Bayesian inference. The two figures on the left and right at the bottom show the data point sequence $y_n=ax_n+b+\varepsilon_n$, the base linear function $y_{Base}=ax+b$ used when the data point sequence was generated, the interpolation result by Bayesian inference $y_{Bayes}=a_{Bayes}x+b_{Bayes}$, and the interpolation result $y_{LSM}=a_{LSM}x+b_{LSM}$ by the least-squares method. It can be seen that there is not much difference between y_{Base}, y_{Bayes}, and y_{LSM}. However, the difference among y_{Base}, y_{Bayes}, and y_{LSM} is a little bit smaller when the x-coordinate sequence is regular than irregular.

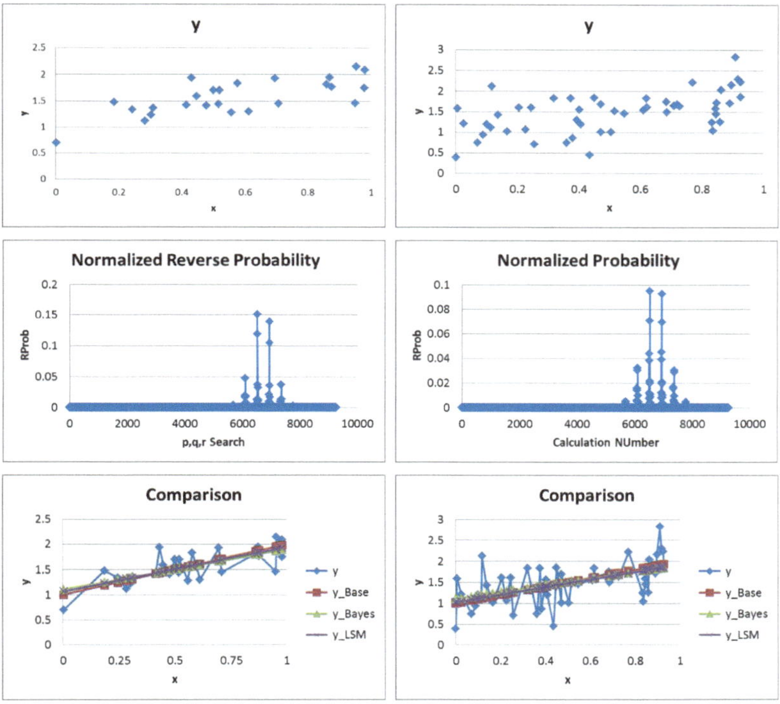

(a) N=25 (b) N=50

Fig. 4.4 A numerical example of interpolation by a linear function (x-direction is a irregular point sequence).

(C) Numerical example 2: Continuous approach; Search for the maximum reverse probability by the steepest ascent method

In (B), the search for the discrete reverse probability maximum was described. Here, a continuous approach will be described. As in the discrete case, the prior probabilities are all equal, respectively. Namely

$$P(a) = const, \text{ for } \forall a;\ P(b) = const, \text{ for } \forall b;\ P(\sigma) = const \text{ for } \forall \sigma. \quad (4.14)$$

Furthermore, the denominator of the reverse probability has nothing to do with the relative magnitude, so it is ignored. At this time, equation (4.6) is derived from equations (4.4) and (4.5) as

$$P(a,b,\sigma \mid \varepsilon_1,\varepsilon_2,\cdots,\varepsilon_N) \sim P(\varepsilon_1,\varepsilon_2,\cdots,\varepsilon_N \mid a,b,\sigma)$$
$$= \prod_{n=1}^{N} P(\varepsilon_n \mid a,b,\sigma) = \prod_{n=1}^{N} \frac{1}{\sqrt{2\pi\sigma^2}} \exp\left(-\frac{(y_n - ax_n - b)^2}{2\sigma^2} \right). \quad (4.15)$$

Since the logarithmic function log is a monotonically increasing function, the relative magnitude does not change even if the logarithm is taken. Therefore, if we take the natural logarithm log to simplify the calculation, we have

$$\log P(a,b,\sigma \mid \varepsilon_1,\varepsilon_2,\cdots,\varepsilon_N) \sim \sum_{n=1}^{N} \left[-\log \frac{1}{\sqrt{2\pi\sigma^2}} - \frac{(y_n - ax_n - b)^2}{2\sigma^2} \right] \quad (4.16)$$

If we take the derivatives with respect to the parameters, we obtain

$$\frac{\partial}{\partial a} \log P(a,b,\sigma \mid \varepsilon_1,\varepsilon_2,\cdots,\varepsilon_N) \sim \sum_{n=1}^{N} \left[\frac{x_n (y_n - ax_n - b)}{\sigma^2} \right], \quad (4.17a)$$

$$\frac{\partial}{\partial b} \log P(a,b,\sigma \mid \varepsilon_1,\varepsilon_2,\cdots,\varepsilon_N) \sim \sum_{n=1}^{N} \left[\frac{(y_n - ax_n - b)}{\sigma^2} \right], \quad (4.17b)$$

$$\frac{\partial}{\partial \sigma^2} \log P(a,b,\sigma \mid \varepsilon_1,\varepsilon_2,\cdots,\varepsilon_N) \sim \sum_{n=1}^{N} \left[-\frac{1}{2\sigma^2} + \frac{(y_n - ax_n - b)^2}{2\sigma^4} \right]. \quad (4.17c)$$

The derivatives when not taking the logarithm are given by

$$\frac{\partial}{\partial a} P(a,b,\sigma \mid \varepsilon_1,\varepsilon_2,\cdots,\varepsilon_N) \sim P(a,b,\sigma \mid \varepsilon_1,\varepsilon_2,\cdots,\varepsilon_N) \sum_{n=1}^{N} \left[\frac{x_n (y_n - ax_n - b)}{\sigma^2} \right], \quad (4.18a)$$

$$\frac{\partial}{\partial b} P(a,b,\sigma \mid \varepsilon_1,\varepsilon_2,\cdots,\varepsilon_N) \sim P(a,b,\sigma \mid \varepsilon_1,\varepsilon_2,\cdots,\varepsilon_N) \sum_{n=1}^{N} \left[\frac{(y_n - ax_n - b)}{\sigma^2} \right], \quad (4.18b)$$

$$\frac{\partial}{\partial \sigma^2} P(a,b,\sigma \mid \varepsilon_1,\varepsilon_2,\cdots,\varepsilon_N) \sim P(a,b,\sigma \mid \varepsilon_1,\varepsilon_2,\cdots,\varepsilon_N) \sum_{n=1}^{N} \left[-\frac{1}{2\sigma^2} + \frac{(y_n - ax_n - b)^2}{2\sigma^4} \right]. \quad (4.18c)$$

In order to verify the above theory, a numerical calculation is performed to search for the maximum value of the numerator of $P_{Numerator}(a,b,\sigma \mid \varepsilon_1,\varepsilon_2,\cdots,\varepsilon_N)$ of $P(a,b,\sigma \mid \varepsilon_1,\varepsilon_2,\cdots,\varepsilon_N)$.

The experimental data used in (B) is used. The following example is the case of $N = 50$. The starting value of the parameter is $a_{start} = 0.5, b_{start} = 0.5, \sigma_{start} = 0.5$. The search step is $\lambda = 0.0002$. The same value $d\lambda = 0.0001$ was used for all of a, b, σ for the numerical differentiation given by

$$\frac{\partial \log P_{Numerator}(a,b,\sigma | \varepsilon_1, \varepsilon_2, \cdots, \varepsilon_N)}{\partial a} = \frac{\log P_{Numerator}(a+d\lambda,b,\sigma | \varepsilon_1, \varepsilon_2, \cdots, \varepsilon_N) - \log P_{Numerator}(a,b,\sigma | \varepsilon_1, \varepsilon_2, \cdots, \varepsilon_N)}{d\lambda}. \tag{4.19}$$

A regular sequence of points is used in the x-direction. Similar results are obtained by using an irregular sequence of points in the x-direction. As shown in Fig. 4.5, it converges smoothly to the maximum value, but in this example, it converges faster when using the numerical derivative.

The programming code is shown in Appendix 4A.

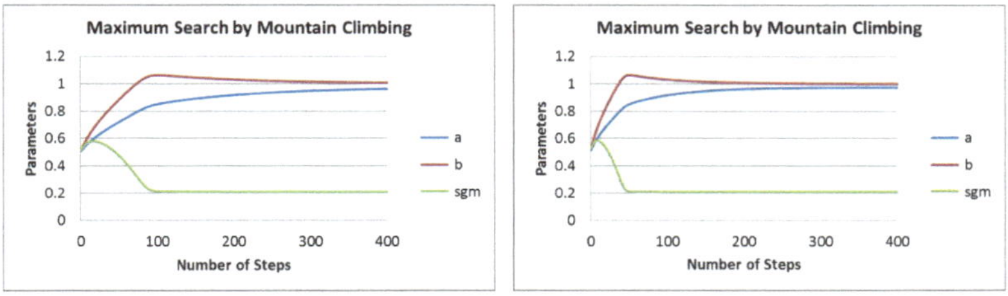

When using analytic differentiation   When using numerical differentiation

Fig. 4.5 Search result of maximum reverse probability by mountain climbing method.

4.1.2. Sinusoidal function

(A) Theory

We consider a problem given by

$$y_n = A\sin(\omega t_n + \phi) + \varepsilon_n, \quad n = 1, 2, \cdots, N \tag{4.20}$$

instead of equation (4.2). We consider a random number ε_n with an average of 0 and a standard deviation of σ following a normal distribution. Namely,

$$\varepsilon_n = y_n - A\sin(\omega t_n + \phi) \sim \frac{1}{\sqrt{2\pi\sigma^2}} \exp\left(-\frac{\varepsilon_n^{\ 2}}{2\sigma^2} \right). \tag{4.21}$$

If we consider A, ω, ϕ as a discrete value $A_i, \omega_j, \phi_k, (i = 1,2.\cdots, I;\ j = 1,2.\cdots, J;\ k = 1,2.\cdots, K)$ instead of a continuous value in a numerical calculation, then equation (4.6) is given by

$$P(A_i,\omega_j,\phi_k \mid \varepsilon_1,\varepsilon_2,\cdots,\varepsilon_N)=\frac{P(\varepsilon_1,\varepsilon_2,\cdots,\varepsilon_N \mid A_i,\omega_j,\phi_k)P(A_i)P(\omega_j)P(\phi_k)}{\sum_i^I\sum_j^J\sum_k^K P(\varepsilon_1,\varepsilon_2,\cdots,\varepsilon_N \mid A_i,\omega_j,\phi_k)P(A_i)P(\omega_j)P(\phi_k)}. \tag{4.22}$$

$P(A_i)$, $P(\omega_j)$, and $P(\phi_k)$ are called the prior probability. $P(\varepsilon_1,\varepsilon_2,\cdots,\varepsilon_N \mid A_i,\omega_j,\phi_k)$ and $P(A_i,\omega_j,\phi_k \mid \varepsilon_1,\varepsilon_2,\cdots,\varepsilon_N)$ are the likelihood function, and the posterior probability, respectively.

If all prior probabilities are equal respectively, that is

$$P(A_i)=const,\ \text{for } \forall i;\ P(\omega_j)=const,\ \text{for } \forall j;\ P(\phi_k)=const\ \text{for } \forall k, \tag{4.23}$$

equation (4.7) becomes

$$P(A_i,\omega_j,\phi_k \mid \varepsilon_1,\varepsilon_2,\cdots,\varepsilon_N)=\frac{P(\varepsilon_1,\varepsilon_2,\cdots,\varepsilon_N \mid A_i,\omega_j,\phi_k)}{\sum_i^I\sum_j^J\sum_k^K P(\varepsilon_1,\varepsilon_2,\cdots,\varepsilon_N \mid A_i,\omega_j,\phi_k)}. \tag{4.24}$$

This is nothing but the likelihood method. For the problem of linear interpolation, the assumption in equation (4.23) seems appropriate. Therefore, in the following numerical example, the estimation of A_i, ω_j, ϕ_k is performed using equation (4.24).

(B) Numerical example: A discrete approach

Numerical calculations are performed to verify the above theory. First, create the equivalent of experimental data. Therefore, a random number sequence according to equations (4.20) and (4.21) is generated. In the example below, we use

$$A=1,\ \omega=1.6\,rad/\sec,\ \phi=45\deg,\ \sigma=0.005, 0.05, 0.25, 0.5\ \ N=100. \tag{4.25}$$

The following are candidates used for unknown discrete parameters:

$$\begin{aligned} A_i &= 0.5+(i-1)\times 0.05,\ i=1,2,\cdots,20, \\ \omega_j &= 1.0+(j-1)\times 0.05,\ j=1,2,\cdots,20, \\ \phi_k &= 30.0+(k-1)\times 1.0,\ k=1,2,\cdots,20. \end{aligned} \tag{4.26}$$

In this case, σ is known. The t direction is a regular sequence. Fig. 4.6 shows the sine signal data with noise. Table 4.1 shows the analysis results by Bayesian inference. It can be seen that the frequency and phase are close to the data given by equation (4.25) even if the noise is quite large. Table 4.2 shows the results when σ

is changed between input and judgment. At the time of judgment, it does not seem to affect the result. It seems that it is not necessary to set to an unknown variable simply to judge the maximum reverse probability.

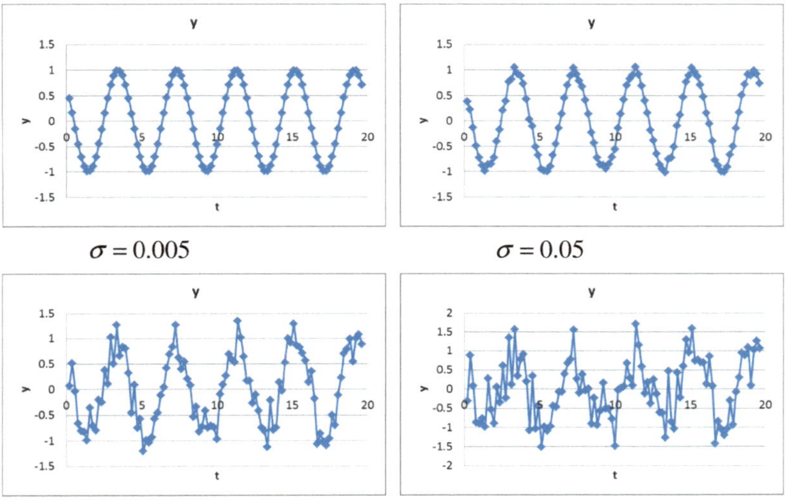

$\sigma = 0.005$ $\sigma = 0.05$

$\sigma = 0.25$ $\sigma = 0.50$

Fig. 4.6 Sinusoidal data with noise

Table 4.1 Sinusoidal signal analysis results 1 (t-direction: regular sequence).

σ	A_{max}	ω_{max}	ϕ_{max}
0.005	1	1.6	45
0.05	1	1.6	45
0.25	0.95	1.6	46
0.5	0.85	1.6	47

Table 4.2 Sinusoidal signal analysis results 2 (t-direction: regular sequence).

for data	σ for judge	A_{max}	ω_{max}	ϕ_{max}
0.2	0.1	0.95	1.6	46
0.2	0.2	0.95	1.6	46
0.2	0.4	0.95	1.6	46

4.2. Identification of a differential equation

(A) Theory

We consider using Bayesian inference to identify differential equations. As a differential equation, we consider the initial value problem of a linear vibration system given by the following linear equation:

$$\frac{d^2\theta}{dt^2} + \nu\frac{d\theta}{dt} + \kappa\theta = 0,\ \theta(0) = 1,\ \dot{\theta}(0) = 0. \tag{4.27}$$

As a result of the experiment, it is assumed that the following data with noise is obtained:

$$y_n = \theta_n + \varepsilon_n. \tag{4.28}$$

This is considered a random sequence with a noise ε_n following a normal distribution with an average of 0 and a standard deviation of σ. In other words

$$\varepsilon_n = y_n - \theta_n \sim \frac{1}{\sqrt{2\pi\sigma^2}}\exp\left(-\frac{\varepsilon_n^{\ 2}}{2\sigma^2}\right). \tag{4.29}$$

In numerical calculation, when ν, κ are considered as discrete values $\nu_i, \kappa_j,\ (i = 1,2.\cdots,I;\ j = 1,2.\cdots,J)$ instead of a continuous value, equation (4.6) becomes

$$P(\nu_i,\kappa_j \,|\, \varepsilon_1,\varepsilon_2,\cdots,\varepsilon_N) = \frac{P(\varepsilon_1,\varepsilon_2,\cdots,\varepsilon_N \,|\, \nu_i,\kappa_j)P(\nu_i)P(\kappa_j)}{\sum_i^I\sum_j^J\sum_k^K P(\varepsilon_1,\varepsilon_2,\cdots,\varepsilon_N \,|\, \nu_i,\kappa_j)P(\nu_i)P(\kappa_j)}. \tag{4.30}$$

As shown in Table 4.2 of section 3.2 (B), the influence of σ is considered to be small and is not set as an unknown parameter. $P(\nu_i)$ and $P(\kappa_j)$ are called the prior probability. $P(\varepsilon_1,\varepsilon_2,\cdots,\varepsilon_N \,|\, \nu_i,\kappa_j)$ and $P(\nu_i,\kappa_j \,|\, \varepsilon_1,\varepsilon_2,\cdots,\varepsilon_N)$ are the likelihood function, and the posterior probability, respectively.

If all prior probabilities are equal respectively, that is

$$P(\nu_i) = const,\ \text{for } \forall i;\ P(\kappa_j) = const,\ \text{for } \forall j, \tag{4.31}$$

equation (4.12) becomes

$$P(\nu_i,\kappa_j \,|\, \varepsilon_1,\varepsilon_2,\cdots,\varepsilon_N) = \frac{P(\varepsilon_1,\varepsilon_2,\cdots,\varepsilon_N \,|\, \nu_i,\kappa_j)}{\sum_i^I\sum_j^J\sum_k^K P(\varepsilon_1,\varepsilon_2,\cdots,\varepsilon_N \,|\, \nu_i,\kappa_j)}. \tag{4.32}$$

This is nothing but the likelihood method. For the problem of linear interpolation, the assumption in equation (4.31) seems appropriate. Therefore, in the following numerical example, the estimation of ν_i, κ_j is performed using equation (4.32).

(B) Numerical examples: A discrete approach

(B-1) Linear case

Numerical calculations are performed to verify the above theory. First, we create the equivalent of experimental data or virtual data. Therefore, a random number sequence according to equations (4.28) and (4.29) is generated. In the example below, we use

$$\nu = 1,\ \kappa = 1.6,\ \phi = 45\,\mathrm{deg},\ \sigma = 0.1,\ N = 50,\ dt = 0.2. \tag{4.33}$$

As a candidate for an unknown discrete parameter ν_i, κ_j, the following is used:

$$\begin{aligned} \nu_i &= i \times 0.02,\ i = 1, 2, \cdots, 10, \\ \kappa_j &= j \times 0.2,\ j = 1, 2, \cdots, 10. \end{aligned} \tag{4.34}$$

In this issue, σ is assumed known. The t-direction is a regular sequence. Fig. 4.7 shows the data y_n with noise. Table 4.3 shows the analysis results by Bayesian inference. It can be seen that the coefficient of the restoration term (natural frequency) is close to the data given by equation (4.33) even if the noise is considerably large.

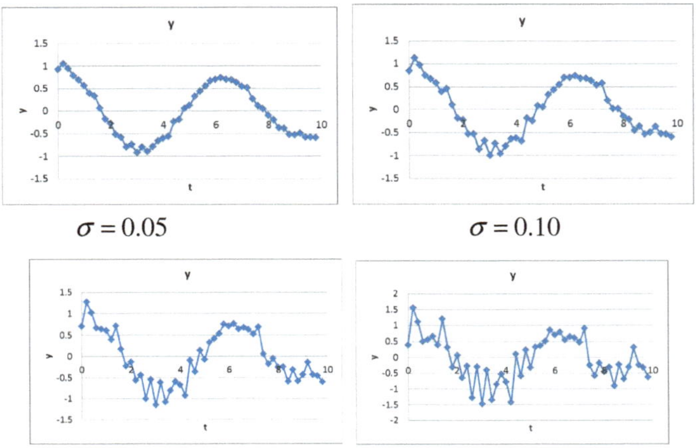

$\sigma = 0.05$ $\sigma = 0.10$

$\sigma = 0.20$ $\sigma = 0.40$

Fig. 4.7 Data with noise

Table 4.3 Analysis results using Bayesian inference (Linear; t -direction: regular sequence).

σ	$\nu_{\max}$	$\kappa_{\max}$
0.05	0.1	1.0
0.1	0.1	1.0
0.2	0.12	1.0
0.4	0.14	1.0

(B-2) Nonlinear case

We consider an example when the restoration term of equation (4.27) becomes non-linear:

$$\frac{d^2\theta}{dt^2}+\nu\frac{d\theta}{dt}+\kappa\sin(\theta)=0,\ \ \theta(0)=1,\ \ \dot{\theta}(0)=0. \tag{4.35}$$

Table 4.4 shows the results of Bayesian inference performed under the same conditions as for the linear case.

Table 4.4 Analysis results using Bayesian inference (Nonlinear; t -direction: regular sequence).

σ	$\nu_{\max}$	$\kappa_{\max}$
0.05	0.1	1.0
0.1	0.12	1.0
0.2	0.14	1.0
0.4	0.16	1.0

With the advent of deep learning, breakthroughs have occurred in AI technologies such as image recognition, voice recognition, and machine translation that humans can easily do but machines cannot. However, deep learning is not the only AI technology. There are problems such as taking a long time to learn and abnormal recognition due to over-fitting, but the biggest problem is anxiety that comes from the fact that the reasoning is a black box and the basis of recognition is completely unknown.

Since Bayesian learning is based on Bayesian inference in statistics, it is based on a theory completely different from deep learning. Since the learning method is completely different, it has the potential to free us from the problems of deep learning described above.

Neural networks are regarded as mathematical models of information processing in the cerebrum. As mentioned in the preceding chapters, the Bayesian inference is also considered to suggest information processing different from neural networks. The Bayesian inference is interesting in that sense.

Learning in the Bayesian inference is quite different from learning in neural networks. Learning in the Bayesian inference is to find the probability of a chosen random variable from a large number of data. In addition, the judgment in the Bayesian inference is to judge what kind of data is aproapriate based on the probability distribution obtained by learning.

In this study, we applied the Bayesian inference to functional interpolation and identification of differential equations. There are traditional solutions to these problems, but we have dealt with what happens from the perspective of the Bayesian inference. I hope that new developments will be brought by looking at it from a new perspective.

REFERENCES 4

[1] Krizhevsky, A., Sutskever, I. & Hinton, G. ImageNet classification with deep convolutional neural networks. In Proc. Advances in Neural Information Processing Systems 25 1090–1098 (2012).

[2] Yann LeCun1, Yoshua Bengio & Geoffrey Hinton, Deep learning, NATURE | VOL 521 | 28 MAY (2015).

[3] Masato Taki, Introduction to Deep Learning, Kodansha (2017) in Japanese.

[4] Nozomu Matsubara, Introduction to Bayesian Statistics, Tokyo Tosho (2008) in Japanese.

[5] Atsushi Suyama, Introduction to Mchine Learning by Bayesian Inference, Kodansha (2017) in Japanese.

[6] Hiroshi Isshiki, Pattern Recognition by Bayesian Inference, The 63rd Joint Meeting of Automatic Control (2020). https://www.jstage.jst.go.jp/article/jacc/63/0/63_329/_pdf

Appendix 4A Code for linear regression

"Microsoft C/C++ Compiler Version 17.00.50727.1 for x86" and "Microsoft Linker Version 11.00.50727" were used for compile and link (in command window; cl source_file_name.c).

(1) Programing code: LinearRegressMtClimbX.c

```
// ---------------------------------------------------------------- //
//                                                                  //
// File Name: LinearRegressMtClimb.c          2021.01.28-2021.02.25 //
// File Name: LinearRegressMtClimbX.c         2021.01.28-2022.05.23 //
//                                                                  //
   //    Linear Regression                                             //
   //                                                                  //
   // ---------------------------------------------------------------- //

   // ----- functions *------------------------------------------------ //

   #include <stdio.h>
   #include <stdlib.h>
   #include <string.h>
   #include <math.h>

   #define PI      3.14159265     // Pi

   void main();
   void pushKey();

   double Uniform( void );                         // uniform random bumber [0,1]
   double rand_normal(double,double);              // normal random number

   double normal(double,double,double);                 // normal distribution
   double Pnml(int, double, double, double, double);    //
   double Pnml_a(int, double, double, double, double);   // differentiation of Pnml with respect
   to a
   double Pnml_b(int, double, double, double, double);   // differentiation of Pnml with respect
   to b
   double Pnml_sgm(int, double, double, double, double); // differentiation of Pnml with respect
   to sgm

   double lognormal(double,double,double);                   // log of normal
   double logPnml(int, double, double, double, double);      // log of Pnml
   double logPnml_a(int, double, double, double, double);   // differentiation of logPnml with
   respect to a
```

```
double logPnml_b(int, double, double, double, double);   // differentiation of logPnml with
respect to b
double logPnml_sgm(int, double, double, double, double); // differentiation of logPnml with
respect to sgm

// ----- variables ------------------------------------------------- //

char title_memo[5000];

int P;                             // number of pattern

int N1;                            // number of sampling

FILE *fp_inp;                      // pointer of input file
FILE *fp_out;                      // pointer of output file

char InputDataFile[80];            // input file name
char OutputDataFile[80];           // output file name

char buf[5000];

double x[10001];                    // x
double y[10001];                    // y ... y = a + b*x + eps
double eps[10001];                  // eps

int x_rand;                         // x is irregular or regular if x_rand = 1 or 0

double a1;                          // a ... given
double b1;                          // b ... given

double lmd;                         // step of mountain climbing
double dlmd;                        // difference in derivative
double a_;                          // parameter y = ax+b
double b_;                          // parameter y = ax+b
double sgm_;                        // parameter stddev of error
double logRProb_;                   // minimum of RProb
double logRProb_a;                  // derivative of RProb by a
double logRProb_b;                  // derivative of RProb by b
double logRProb_sgm;                // derivative of RProb by sgm
```

```
double Value[10001];            // value
double x_max;
double x_min;
double Dx;
double width;

double myu[101];                // average
double sgm[101];                // standard deviation
double myu1;                    // specified
double sgm1;                    // specified

double sgm0;                    // initial value of sgm
double a0;                      // initial value of a
double b0;                      // initial value of b

int ipnt[10001];

// ---------------------------------------------------------------- //

void main()
{
    int p, q, r, m, n;
    int i, j, k;
    double sum, tmp, tmp_a, tmp_b, tmp_sgm;

    ////  Open input file
    sprintf(InputDataFile, "LinearRegressMtClimbX_inp.dat");

    if ((fp_inp = fopen(InputDataFile, "r")) == NULL) {
        printf("Failed in Reading Input Data File! ... %s¥n", InputDataFile);
        exit(1);
    }

    //// Open output file
    sprintf(OutputDataFile, "LinearRegressMtClimbX_out.csv");

    if ((fp_out = fopen(OutputDataFile, "w")) == NULL) {
        printf("Failed in Reading Output Data File! ... %s¥n", OutputDataFile);
```

```
    exit(1);
}

//// input from file
fscanf(fp_inp, "%s", title_memo);

fscanf(fp_inp, "%s %lf", buf, &x_min);
fscanf(fp_inp, "%s %lf", buf, &x_max);
fscanf(fp_inp, "%s %lf", buf, &Dx);

fscanf(fp_inp, "%s %d", buf, &P);
fscanf(fp_inp, "%s %d", buf, &N1);
fscanf(fp_inp, "%s %lf", buf, &myu1);
fscanf(fp_inp, "%s %lf", buf, &sgm1);

fscanf(fp_inp, "%s %lf", buf, &a1);
fscanf(fp_inp, "%s %lf", buf, &b1);

fscanf(fp_inp, "%s %lf", buf, &sgm0);

fscanf(fp_inp, "%s %lf", buf, &a0);
fscanf(fp_inp, "%s %lf", buf, &b0);

fscanf(fp_inp, "%s %lf", buf, &dlmd);
fscanf(fp_inp, "%s %lf", buf, &lmd);


printf("memo: %s¥n", title_memo);
printf("¥n");

printf("x_min = %12.6f¥n", x_min);
printf("x_max = %12.6f¥n", x_max);
printf("Dx    = %12.6f¥n", Dx);
printf("¥n");

printf("P      = %d¥n", P);
printf("N1     = %d¥n", N1);
printf("myu1   = %12.6f¥n", myu1);
printf("sgm1   = %12.6f¥n", sgm1);
```

```
printf("a1     = %12.6f¥n", a1);
printf("b1     = %12.6f¥n", b1);

printf("sgm0   = %12.6f¥n", sgm0);

printf("a0     = %12.6f¥n", a0);
printf("b0     = %12.6f¥n", b0);

printf("x_rand = %d¥n", x_rand);
printf("¥n");

printf("dlmd = %12.6f¥n", dlmd);
printf("lmd  = %12.6f¥n", lmd);
printf("¥n");


fprintf(fp_out, "memo: %s¥n", title_memo);
fprintf(fp_out, "¥n");

fprintf(fp_out, "x_min =, %12.6f¥n", x_min);
fprintf(fp_out, "x_max =, %12.6f¥n", x_max);
fprintf(fp_out, "Dx =, %12.6f¥n", Dx);
fprintf(fp_out, "¥n");

fprintf(fp_out, "P =, %d¥n", P);
fprintf(fp_out, "N1 =, %d¥n", N1);
fprintf(fp_out, "myu1 =, %12.6f¥n", myu1);
fprintf(fp_out, "sgm1 =, %12.6f¥n", sgm1);

fprintf(fp_out, "a1 =, %12.6f¥n", a1);
fprintf(fp_out, "b1 =, %12.6f¥n", b1);

fprintf(fp_out, "sgm1 =, %12.6f¥n", sgm0);

fprintf(fp_out, "a1 =, %12.6f¥n", a0);
fprintf(fp_out, "b1 =, %12.6f¥n", b0);

fprintf(fp_out, "dlmd =, %12.6f¥n", dlmd);
```

```
fprintf(fp_out, "lmd =, %12.6f¥n", lmd);
fprintf(fp_out, "¥n");

width = x_max-x_min;

// input data

for (n = 1; n <= N1; n++)
    Value[n] = rand_normal(myu1,sgm1);

 // y[n] = a1*x[n]+b1+eps[n]

for (n = 1; n <= N1; n++)
    eps[n] = Value[n];

for (n = 1; n <= N1; n++)
    x[n] = (n+0.0)*1.0/(N1+0.0);

for (n = 1; n <= N1; n++)
    y[n] = a1*x[n] + b1 + eps[n];

fprintf(fp_out, "n, x, y¥n");
for (n = 1; n <= N1; n++)
    fprintf(fp_out, "%d, %12.6f, %12.6f¥n", n, x[n], y[n]);
fprintf(fp_out, "¥n");


// Mountain Climbing ... analytical differentiation

fprintf(fp_out, "Mountain Climbing ... analytical differentiation¥n");
a_ = a0;
b_ = b0;
sgm_ = sgm0;
logRProb_ = 0.0;
for (n = 1; n <= N1; n++)
    logRProb_ += logPnml(n, a_, b_, myu1, sgm_);

fprintf(fp_out, "i, a, b, sgm, logRProb, logRProb_a, logRProb_b, logRProb_sgm¥n");
for (i = 1; i <= 500; i++) {
```

```
    logRProb_a = 0.0;
    logRProb_b = 0.0;
    logRProb_sgm = 0.0;
    tmp_a = 0.0;
    tmp_b = 0.0;
    tmp_sgm = 0.0;
    for (n = 1; n <= N1; n++) {
        tmp_a += logPnml_a(n, a_, b_, myu1, sgm_);
        tmp_b += logPnml_b(n, a_, b_, myu1, sgm_);
        tmp_sgm += logPnml_sgm(n, a_, b_, myu1, sgm_);
    }
    logRProb_a += tmp_a;
    logRProb_b += tmp_b;
    logRProb_sgm += tmp_sgm;

    a_ += logRProb_a*dlmd;
    b_ += logRProb_b*dlmd;
    sgm_ += logRProb_sgm*dlmd;

    logRProb_ = 0.0;
    for (n = 1; n <= N1; n++)
        logRProb_ += logPnml(n, a_, b_, myu1, sgm_);
    fprintf(fp_out, "%d, %12.6f, %12.6f, %12.6f, %12.6f, %12.6f, %12.6f, %12.6f¥n",
            i, a_, b_, sgm_, logRProb_, logRProb_a, logRProb_b, logRProb_sgm);
}
fprintf(fp_out, "¥n");


// Mountain Climbing ... numerical differentiation

fprintf(fp_out, "Mountain Climbing ... numerical differentiation¥n");
a_ = a0;
b_ = b0;
sgm_ = sgm0;
logRProb_ = 0.0;
for (n = 1; n <= N1; n++)
    logRProb_ += logPnml(n, a_, b_, myu1, sgm_);

fprintf(fp_out, "i, a, b, sgm, logRProb, logRProb_a, logRProb_b, logRProb_sgm¥n");
```

```
        for (i = 1; i <= 500; i++) {

            logRProb_a = 0.0;
            for (n = 1; n <= N1; n++)
                logRProb_a += logPnml(n, a_+dlmd, b_, myu1, sgm_);
            logRProb_a = (logRProb_a-logRProb_)/dlmd;

            logRProb_b = 0.0;
            for (n = 1; n <= N1; n++)
                logRProb_b += logPnml(n, a_, b_+dlmd, myu1, sgm_);
            logRProb_b = (logRProb_b-logRProb_)/dlmd;

            logRProb_sgm = 0.0;
            for (n = 1; n <= N1; n++)
                logRProb_sgm += logPnml(n, a_, b_, myu1, sgm_+dlmd);
            logRProb_sgm = (logRProb_sgm-logRProb_)/dlmd;

            a_ += logRProb_a*lmd;
            b_ += logRProb_b*lmd;
            sgm_ += logRProb_sgm*lmd;

            logRProb_ = 0.0;
            for (n = 1; n <= N1; n++)
                logRProb_ += logPnml(n, a_, b_, myu1, sgm_);
            fprintf(fp_out, "%d, %12.6f, %12.6f, %12.6f, %12.6f, %12.6f, %12.6f, %12.6f¥n",
                    i, a_, b_, sgm_, logRProb_, logRProb_a, logRProb_b, logRProb_sgm);
        }
        fprintf(fp_out, "¥n");


    fprintf(fp_out, "¥n");
    fclose(fp_out);

    pushKey();
}

// ------------------------------------------------------------------ //

void pushKey()
```

```
{
    printf("¥n      Push Return Key! ");
    getchar();
    getchar();
}

// ------------------------------------------------------------ //

double rand_normal( double myu, double sigma )
{
    double z=sqrt( -2.0*log(Uniform()) ) * sin( 2.0*PI*Uniform() );

    return myu + sigma*z;
 }

// ------------------------------------------------------------ //

double Uniform( void )
{
    static int x=10;
    int a=1103515245,b=12345,c=2147483647;
    x = (a*x + b)&c;

    return ((double)x+1.0) / ((double)c+2.0);
}

// ------------------------------------------------------------ //


double normal(double x, double myu, double sgm)     // normal distribution
{
    return 1.0/sqrt(2.0*PI*sgm*sgm)*exp(-(x-myu)*(x-myu)/2.0/sgm/sgm);
}

// ------------------------------------------------------------ //

double Pnml(int n, double a, double b, double myu, double sgm)
{
    return normal(y[n]-a*x[n]-b, myu, sgm);
```

```
}

// ------------------------------------------------------------ //

double Pnml_a(int n, double a, double b, double myu, double sgm)
{
    return normal(y[n]-a*x[n]-b, myu, sgm)
          *x[n]*(y[n]-a*x[n]-b)/sgm/sgm;
}

// ------------------------------------------------------------ //

double Pnml_b(int n, double a, double b, double myu, double sgm)
{
    return normal(y[n]-a*x[n]-b, myu, sgm)
          *(y[n]-a*x[n]-b)/sgm/sgm;
}

// ------------------------------------------------------------ //

double Pnml_sgm(int n, double a, double b, double myu, double sgm)
{
    return ( -0.5/sgm/sgm*normal(y[n]-a*x[n]-b, myu, sgm)
            +normal(y[n]-a*x[n]-b, myu,
sgm)*(y[n]-a*x[n]-b)*(y[n]-a*x[n]-b)/2.0/sgm/sgm/sgm/sgm)*2.0*sgm;
}

// ------------------------------------------------------------ //

double lognormal(double x, double myu, double sgm)      // log of normal disribution
{
    return log(1.0/sqrt(2.0*PI*sgm*sgm)) - (x-myu)*(x-myu)/2.0/sgm/sgm;
}

// ------------------------------------------------------------ //

double logPnml(int n, double a, double b, double myu, double sgm)
{
    return lognormal(y[n]-a*x[n]-b, myu, sgm);
```

```
}

// ------------------------------------------------------------------ //

double logPnml_a(int n, double a, double b, double myu, double sgm)
{
    return x[n]*(y[n]-a*x[n]-b)/sgm/sgm;
}

// ------------------------------------------------------------------ //

double logPnml_b(int n, double a, double b, double myu, double sgm)
{
    return (y[n]-a*x[n]-b)/sgm/sgm;
}

// ------------------------------------------------------------------ //

double logPnml_sgm(int n, double a, double b, double myu, double sgm)
{
    return (-1.0/2.0/sgm/sgm + (y[n]-a*x[n]-b)*(y[n]-a*x[n]-b)/2.0/sgm/sgm/sgm/sgm )*2.0*sgm;
}

// ------------------------------------------------------------------ //
```

(2) Input file: LinearRegressMtClimbX_inp.dat

```
ProbDistParam_20230523
x_min        -1.5
x_max        +1.5
Dx           0.25
P            20
N1           50
myu1         0.0
sgm1         0.2
a1           1.0
b1           1.0
sgm0         0.5
a0           0.5
```

b0	0.5
dlmd	0.0001
lmd	0.0002

5. FUNDAMENTALS OF DATA ASSIMILATION

Numerical simulation technology has made remarkable progress. Calculation accuracy, speed, and scale of the target have improved dramatically. Complex fluid motion, combustion problems so on can be calculated using state-of-the-art supercomputers.

On the other hand, the limits of numerical simulation have become clear. If the object is not captured in a completely explicit form, problems such as earthquake prediction, for example, could not be dealt with satisfactorily. At the present point, it will be difficult to grasp the structure of the crust with sufficient accuracy. Therefore, it will be difficult to accurately replace the motion as a mathematical problem. Even in such a case, the situation will be different if there is something to supplement the unknown part, for example, observation data. Earthquake prediction is still difficult, but depending on the problem, by combining observation data with numerical simulation, it would be possible to grasp the phenomenon with considerable accuracy.

In the first place, the state space model created in the control world was applied to improve the accuracy of meteorological forecasting [1-2]. Current meteorological forecasts combine large-scale numerical calculations using supercomputers with global meteorological observations. As a result, the accuracy of weather forecasts has improved surprisingly.

After that, the effectiveness of combining data observation and numerical simulation was recognized, and it came to be called data assimilation technology, and its application in various fields became active [3-4]. However, its application is not widespread, it is still in the early stages of development, and social awareness of this technology is low.

It is a technology for learning using data, but the difference from neural networks is that it uses a mathematical model that expresses phenomena as its skeleton. A neural network is a method of learning using only data without a model. Since data assimilation is a method of learning the unknown part of a mathematical model with data, it enables efficient and highly accurate inference.

Therefore, knowledge of phenomena and their mathematics and knowledge of statistics that are the basis of data assimilation, especially Bayesian statistics, are required. Especially the part related to Bayesian statistics is advanced and not easy to understand. We hope that the explanations in this chapter will help beginners understanding.

5.1. Basic concept of data assimilation

5.1.1. Problem setting

In the following discussion, the process of the time evolution of the system is known, but there are unknown parts and it is difficult to grasp it accurately and quantitatively. However, it is assumed that the observation data of the system can be obtained to supplement it. What we discuss in this chapter is how to get a quantitative grasp of the processes taking place in the system in such cases. Therefore, it is a major premise that the process is known and that there is observational data.

Neural networks, on the other hand, are discussions when the process is unknown. In data assimilation, the skeleton of the problem is known as a system model. Since the purpose of this section is a basic consideration of data assimilation, we will leave the full-scale discussion to the next section, and in this section, we will discuss a method that does not make full use of Bayesian statistics.

Consider the case where both the state vector x and the observation vector y are one-dimensional as shown below. Using the time $t = 1, 2, \cdots, T$, the system model with the noise v and the observation model with the noise w are given by

$$x_t = Ax_{t-1} + u_t + v_t, \quad t = 1, 2, \cdots, T \tag{5.1}$$

and

$$y_t = Bx_t + w_t, \quad t = 1, 2, \cdots, T, \tag{5.2}$$

respectively. Since the system noise v is not essential, it can be set to 0. A and B are constants, u is an external input, v and w are normal random numbers with variances σ_v^2 and σ_w^2:

$$p(v_t) = \frac{1}{\sqrt{2\pi\sigma_v^2}} \exp\left(-\frac{v_t^2}{2\sigma_v^2}\right), \tag{5.3}$$

$$p(w_t) = \frac{1}{\sqrt{2\pi\sigma_w^2}} \exp\left(-\frac{w_t^2}{2\sigma_w^2}\right). \tag{5.4}$$

5.1.2. Numerical results

Since variance σ_v^2 is not essential, we assume $\sigma_v^2 = 0$ in the following calculations. We consider a virtual problem for the data assimilation. Let time be $t = 1, 2, \cdots, T$, an initial value of state variable be $x_1 = 0.1$, a parameter be $A = 1$, and the external input be

$$u_t = 0.05 \sin\left(\frac{2\pi}{30} t\right). \tag{5.5}$$

Furthermore, assuming the parameter for the system model as $B = 1$ and the variance of the system noise as $\sigma_w^2 = 0.05^2$, we generate a virtual system state x and virtual observation data y.

Consider the problem of estimating the parameters A and B by assuming the observation data y known. That is, for the assumed parameters $\hat{A}$ and $\hat{B}$, the assumed state x is obtained by equation (5.1), and the noise y - Bx is then obtained by equation (5.2). From equation (5.2), $\hat{A}$ and $\hat{B}$ making y - Bx as white noise are considered to be the estimated values of the parameters A and B.

(a) Method using least square method (LSM)

The simplest way of data assimilation is to apply the least square method (LSM):

$$S(\hat{A},\hat{B}) = \sum_{t=1}^{T}(y_t - \hat{B}\hat{x}_t)^2 = \min. \tag{5.6}$$

$\hat{A}$ and $\hat{B}$ can be determined by LSM. The parameter $\hat{A}$ is hidden in $\hat{x}$. We calculate the derivatives of S using the numerical differentiation:

$$\frac{\partial S}{\partial \hat{A}} \approx \frac{1}{\lambda}\left(S(\hat{A}+d\lambda,\hat{B}) - S(\hat{A},\hat{B})\right),\quad \frac{\partial S}{\partial \hat{B}} \approx \frac{1}{\lambda}\left(S(\hat{A},\hat{B}+d\lambda) - S(\hat{A},\hat{B})\right). \tag{5.7}$$

For the estimation of $\hat{A}$ and $\hat{B}$, the steepest decent method is applied:

$$\hat{A} \leftarrow \hat{A} - \frac{\partial S}{\partial \lambda}dl,\quad \hat{B} \leftarrow \hat{B} - \frac{\partial S}{\partial \lambda}dl. \tag{5.8}$$

The numerical results with $d\lambda = 1\times10^{-7}$ and $dl = 2\times10^{-6}$ are shown in Fig. 5.1 and Fig. 5.2. Fig. 5.1 shows that the unknown parameters $\hat{A}$ and $\hat{B}$ converge to the set values of 1 and 1 from 0.3 and 0.5, respectively. Fig. 5.2 Shows that the initial state x_{ini} converges to the data assimilated state x.

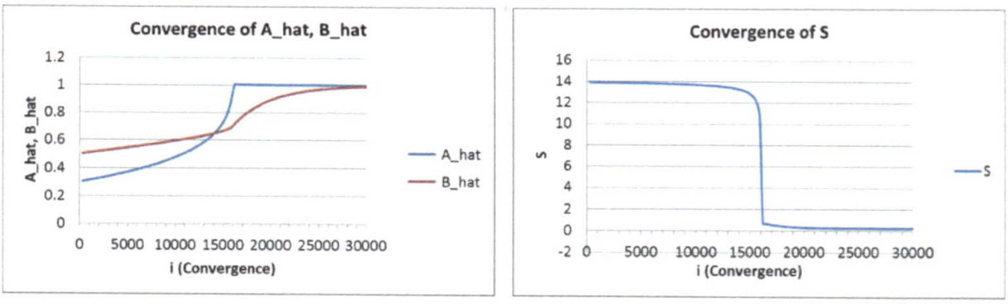

Fig. 5.1. Convergence of Steepest Ascent Method in Search of Solution

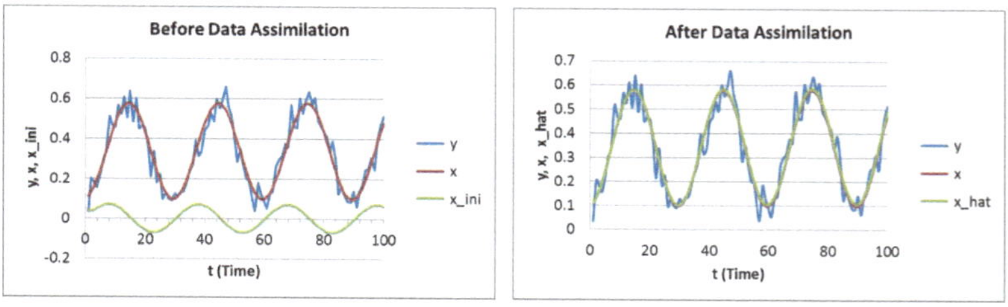

Fig. 5.2. Solution by Data Assimilation (Least Square Method)

(b) Method using maximum likelihood

If we use that $y_t - Bx_t$ is Gaussian noise with the variance $\sigma_w{}^2$:

$$p(w_t) = p(y_t - Bx_t) = \frac{1}{\sqrt{2\pi\sigma_w{}^2}} \exp\left(-\frac{(y_t - Bx_t)^2}{2\sigma_w{}^2} \right), \tag{5.9}$$

we can estimate the parameters A and B. Namely, since the joint probability of $y_t - Bx_t$, $t = 1, 2, \cdots, T$ becomes

$$\begin{aligned} P(\hat{A}, \hat{B}) &= p(y_1 - \hat{B}x_1, y_2 - \hat{B}x_2, \cdots, y_T - \hat{B}x_T) \\ &= \prod_{t=1}^{T} p(y_t - \hat{B}x_t) = \prod_{t=1}^{T} \frac{1}{\sqrt{2\pi\sigma_w{}^2}} \exp\left(-\frac{(y_t - \hat{B}x_t)^2}{2\sigma_w{}^2} \right), \end{aligned} \tag{5.10}$$

we can estimate A and B by applying the maximum likelihood estimation (MLE). MLE means that the parameters A and B making $P(\hat{A}, \hat{B})$ maximum can give the estimate of A and B. MLE is obtained from Bayes inference under the condition of equal prior probabilities.

The natural logarithm of $P(\hat{A}, \hat{B})$, that is, $\log_e P(\hat{A}, \hat{B})$ is used to avoid the underflow. The maximum of $P(\hat{A}, \hat{B})$ is obtained by the steepest ascent method. The differentiation is obtained by the same numerical differentiation as in equation (5.7), and since it is the steepest ascent method, the minus on the right side of Eq. (5.8) is changed to plus.

The numerical results with $d\lambda = 1 \times 10^7$ and $dl = 1 \times 10^{-6}$ are shown in Fig. 5.3 and Fig. 5.4. In the figures, not only the unknown parameters A and B but also the state x converge to the set value by the data assimilation. If the results by the least square method are compared with those by the maximum likelihood estimation, the convergence of the former is a little bit faster than that of the latter, since $dl = 2 \times 10^{-6}$ is used in the former and $dl = 1 \times 10^{-6}$ for the latter.

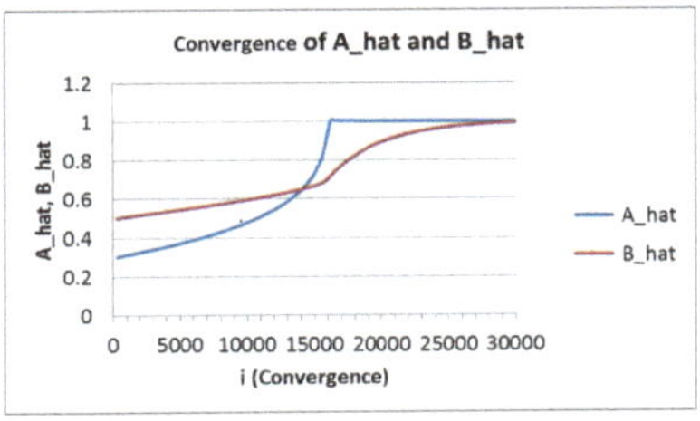

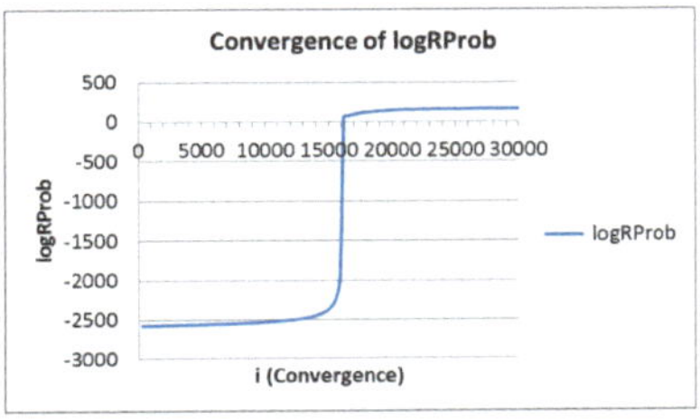

Fig. 5.3. Convergence of Mountain Climb Method in Search of Solution

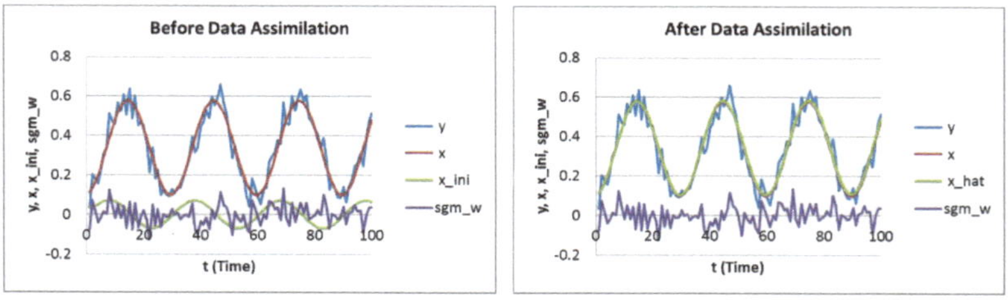

Fig. 5.4. Solution by Data Assimilation (Maximum Likelihood Estimation)

5.2. Sequential estimation using Bayesian inference and state space model

There is a system and we consider the time evolution of the process. Let t be the time, $\mathbf{x}$ be the state vector, and $\mathbf{v}$ be the system noise, and the process model is given below:

$$\mathbf{x}_t = F(\mathbf{x}_{t-1}) + \mathbf{v}_t . \tag{5.11}$$

Since the noise $\mathbf{v}$ is not essential, it could be 0.

On the other hand, we suppose there is system observation data. Let $\mathbf{y}$ be the observation vector and $\mathbf{w}$ be the observation noise, and the observation model is given as follows:

$$\mathbf{y}_t = H(\mathbf{x}_t) + \mathbf{w}_t . \tag{5.12}$$

The dimension of the observation vector is less than or equal to the dimension of the state vector.

There are many possible solutions to this problem. As a non-sequential method, if we consider a method of solving each block of data, there is the least square method or the maximum likelihood estimation method. Discussed below is a completely sequential solution using Bayesian inference. First, Bayes' theorem will be described.

When the cause is x and the result is y, the following equation holds for the joint probability $p(x, y)$:

$$p(x, y) = p(x \mid y)\,p(y) = p(y \mid x)\,p(x) . \tag{5.13}$$

From this, Bayes' theorem is given below:

$$p(x \mid y) = \frac{p(y \mid x)\,p(x)}{p(y)} , \tag{5.14}$$

where $p(x)$, $p(y \mid x)$, and $p(x \mid y)$ are called as prior probability, likelihood function, and posterior probability, respectively. The likelihood function $p(y \mid x)$ gives the probability of a result y with respect to a cause x. On the other hand, $p(x \mid y)$ is the probability of cause x when the result is y. Considering the result y fixedly, it can be inferred that the x that maximizes $p(x|y)$ among the possible values of the cause x is the cause x that brought about the specific result y. Bayes' theorem makes such inference possible. This inference is called Bayesian inference.

The conditional probability of $\mathbf{x}_t$ on condition $\mathbf{y}_t$ is given by

$$p(\mathbf{x}_t \mid \mathbf{y}_{1:t}) = \frac{p(\mathbf{y}_t \mid \mathbf{x}_t) p(\mathbf{x}_t \mid \mathbf{y}_{1:t-1})}{p(\mathbf{y}_t \mid \mathbf{y}_{1:t-1})}, \tag{5.15}$$

where

$$\mathbf{y}_{1:t-1} = (\mathbf{y}_{t-1}, \mathbf{y}_{t-2}, \cdots, \mathbf{y}_1). \tag{5.16}$$

Equation (5.15) is derived as follows. Since the transition $\mathbf{y}_{t-1} \to (\mathbf{y}_t, \mathbf{x}_t)$ occurs with $\mathbf{y}_{t-1} \to \mathbf{x}_t \to \mathbf{y}_t$, we have

$$\begin{aligned} p(\mathbf{x}_t \mid \mathbf{y}_{1:t}) &= \frac{p(\mathbf{x}_t, \mathbf{y}_{1:t})}{p(\mathbf{y}_{1:t})} = \frac{p(\mathbf{y}_t, \mathbf{y}_{1:t-1}, \mathbf{x}_t)}{p(\mathbf{y}_t, \mathbf{y}_{1:t-1})} = \frac{p(\mathbf{y}_t, \mathbf{x}_t \mid \mathbf{y}_{1:t-1}) p(\mathbf{y}_{1:t-1})}{p(\mathbf{y}_t \mid \mathbf{y}_{1:t-1}) p(\mathbf{y}_{1:t-1})} \\ &= \frac{p(\mathbf{y}_t, \mathbf{x}_t \mid \mathbf{y}_{1:t-1})}{p(\mathbf{y}_t \mid \mathbf{y}_{1:t-1})} = \frac{p(\mathbf{y}_t \mid \mathbf{x}_t) p(\mathbf{x}_t \mid \mathbf{y}_{1:t-1})}{p(\mathbf{y}_t \mid \mathbf{y}_{1:t-1})}. \end{aligned} \tag{5.17}$$

The denominator is constant, and we then obtain

$$p(\mathbf{x}_t \mid \mathbf{y}_{1:t}) \propto p(\mathbf{y}_t \mid \mathbf{x}_t) p(\mathbf{x}_t \mid \mathbf{y}_{1:t-1}). \tag{5.18}$$

$p(\mathbf{y}_t \mid \mathbf{x}_t)$ is obtained from the observation equation. Namely, the mean of $\mathbf{y}_t$ is $H(\mathbf{x}_t)$, and the covariance is calculated from $\mathbf{w}_t$. $p(\mathbf{x}_t \mid \mathbf{y}_{1:t-1})$ is given by

$$p(\mathbf{x}_t \mid \mathbf{y}_{1:t-1}) = \int p(\mathbf{x}_t \mid \mathbf{x}_{t-1}) p(\mathbf{x}_{t-1} \mid \mathbf{y}_{1:t-1}) d\mathbf{x}_{t-1}. \tag{5.19}$$

$p(\mathbf{x}_t \mid \mathbf{x}_{t-1})$ is obtained from the system equation. The denominator $p(\mathbf{y}_t \mid \mathbf{y}_{1:t-1})$ of Bayes' theorem given by equation (5.15) is the integral of the numerator $p(\mathbf{y}_t \mid \mathbf{x}_t) p(\mathbf{x}_t \mid \mathbf{y}_{1:t-1})$ with respect to $\mathbf{x}_t$. Therefore, at time t, since the probability distributions $p(\mathbf{y}_t \mid \mathbf{x}_t)$ and $p(\mathbf{x}_t \mid \mathbf{x}_{t-1})$ can be calculated from the observation equation and the system equation, if $p(\mathbf{x}_{t-1} \mid \mathbf{y}_{t-1})$ can be calculated, $p(\mathbf{x}_t \mid \mathbf{y}_{1:t-1})$ can be calculated from equation (5.19), and $p(\mathbf{x}_t \mid \mathbf{y}_{1:t})$ can be calculated from equation (5.15). This means that Bayesian probabilities can be calculated sequentially over time given the initial conditions:

Initial value: $p(\mathbf{x}_0 \mid \mathbf{y}_0) \to p(\mathbf{x}_1 \mid \mathbf{y}_{1:0})$ from equation (5.19) and $p(\mathbf{y}_1 \mid \mathbf{y}_{1:0})$ $\to p(\mathbf{x}_1 \mid \mathbf{y}_{1:1})$ from equation (5.15) $\to p(\mathbf{x}_2 \mid \mathbf{y}_{1:1})$ from equation (5.19) and $p(\mathbf{y}_2 \mid \mathbf{y}_{1:1})$ $\to p(\mathbf{x}_2 \mid \mathbf{y}_{1:2})$ from equation (5.15) $\to \cdots$

If we know the probability of each time, it means that we know the mean and covariance of $\mathbf{x}_t$ at each time. That is, the answer to the problem is obtained.

5.3. The simplest case

5.3.1. Sequential estimation solution

We apply the theory of sequential estimation described in section 5.2 to the same simple problem considered in section 5.1.2. That is, we consider the case where the state vector is one-dimensional, namely a scalar. The system model is given by

$$x_t = Ax_{t-1} + u_t + v_t, \quad t = 1, 2, \cdots, T, \tag{5.1}$$

and the observation model by

$$y_t = Bx_t + w_t, \quad t = 1, 2, \cdots, T, \tag{5.2}$$

respectively where A and B are constants, and u is an external input. Since the noise v_t is not essential, it can be 0. The noises v_t and w_t follow normal distributions with mean 0 and variances σ_v^2 and σ_w^2:

$$p(v_t) = \frac{1}{\sqrt{2\pi\sigma_v^2}} \exp\left(-\frac{v_t^2}{2\sigma_v^2} \right), \tag{5.3}$$

$$p(w_t) = \frac{1}{\sqrt{2\pi\sigma_w^2}} \exp\left(-\frac{w_t^2}{2\sigma_w^2} \right). \tag{5.4}$$

If the noise v_t is equal to 0, $p(v_t)$ becomes

$$p(v_t) = \lim_{\sigma_v \to 0} \frac{1}{\sqrt{2\pi\sigma_v^2}} \exp\left(-\frac{v_t^2}{2\sigma_v^2} \right) = \delta(v_t), \tag{5.20}$$

where $\delta(v_t)$ is Dirac's delta function.

Since the conditional probability $p(x_t \mid y_{1:t-1})$ given by equation (5.19) is rewritten as

$$\begin{aligned} p(x_t \mid y_{1:t-1}) &= \int_{-\infty}^{\infty} p(x_t \mid x_{t-1}) p(x_{t-1} \mid y_{1:t-1}) dx_{t-1} \\ &= \int_{-\infty}^{\infty} \frac{1}{\sqrt{2\pi\sigma_v^2}} \exp\left(-\frac{(x_t - Ax_{t-1} - u_t)^2}{2\sigma_v^2} \right) p(x_{t-1} \mid y_{1:t-1}) dx_{t-1}, \end{aligned} \tag{5.21}$$

we can calculate it, if we know $p(x_{t-1} \mid y_{1:t-1})$. When $y_{1:t-1}$ is known, x_{t-1} is fixed as $x_{t-1|t-1}$. If we assume $p(x_{t-1} \mid y_{1:t-1}) = p(x_{t-1} \mid x_{t-1|t-1})$ follows a normal distribution:

$$p(x_{t-1} \mid y_{1:t-1}) = p(x_{t-1} \mid x_{t-1|t-1}) = \frac{1}{\sqrt{2\pi\sigma_{t-1|t-1}^2}} \exp\left(-\frac{(x_{t-1} - x_{t-1|t-1})^2}{2\sigma_{t-1|t-1}^2} \right), \tag{5.22}$$

and substituting this into equation (5.21) (→Appendix 5A), we obtain

$$p(x_t \mid y_{1:t-1}) = \frac{1}{2\pi\sigma_v \sigma_{t-1|t-1}} \int_{-\infty}^{\infty} \exp\left(-\frac{(x_t - Ax_{t-1} - u_t)^2}{2\sigma_v^{\ 2}} - \frac{(x_{t-1} - x_{t-1|t-1})^2}{2\sigma_{t-1|t-1}^{\ 2}} \right) dx_{t-1}$$

$$= \frac{1}{2\pi\sigma_v \sigma_{t-1|t-1}} \int_{-\infty}^{\infty} \exp\left(-\frac{\sigma_v^{\ 2} + \sigma_{t-1|t-1}^{\ 2} A^2}{2\sigma_v^{\ 2} \sigma_{t-1|t-1}^{\ 2}} \left[(x_{t-1} - x_{t-1|t-1}) - \left(\frac{\sigma_{t-1|t-1}^{\ 2} A(x_t - Ax_{t-1|t-1} - u_t)}{\sigma_v^{\ 2} + \sigma_{t-1|t-1}^{\ 2} A^2} \right) \right]^2 \right) dx_{t-1} \cdot$$

$$\cdot \exp\left(-\frac{(x_t - Ax_{t-1|t-1} - u_t)^2}{2(\sigma_v^{\ 2} + \sigma_{t-1|t-1}^{\ 2} A^2)} \right). \tag{5.23}$$

Since we have

$$\int_{-\infty}^{\infty} \exp\left(-\frac{\sigma_v^{\ 2} + \sigma_{t-1|t-1}^{\ 2} A^2}{2\sigma_v^{\ 2} \sigma_{t-1|t-1}^{\ 2}} \left[(x_{t-1} - x_{t-1|t-1}) - \left(\frac{\sigma_{t-1|t-1}^{\ 2} A(x_t - Ax_{t-1|t-1} - u_t)}{\sigma_v^{\ 2} + \sigma_{t-1|t-1}^{\ 2} A^2} \right) \right]^2 \right) dx_{t-1}$$

$$= \int_{-\infty}^{\infty} \exp\left(-\frac{\sigma_v^{\ 2} + \sigma_{t-1|t-1}^{\ 2} A^2}{2\sigma_v^{\ 2} \sigma_{t-1|t-1}^{\ 2}} x_{t-1}^{\ 2} \right) dx_{t-1} = \sqrt{2\pi} \frac{\sigma_v \sigma_{t-1|t-1}}{\sqrt{\sigma_v^{\ 2} + \sigma_{t-1|t-1}^{\ 2} A^2}}, \tag{5.24}$$

because of $\int_{-\infty}^{\infty} \exp\{-a(u-b)^2\} dx = \int_{-\infty}^{\infty} \exp(-au^2) dx$, we obtain

$$p(x_t \mid y_{1:t-1}) = \frac{1}{\sqrt{2\pi(\sigma_v^{\ 2} + \sigma_{t-1|t-1}^{\ 2} A^2)}} \exp\left(-\frac{(x_t - Ax_{t-1|t-1} - u_t)^2}{2(\sigma_v^{\ 2} + \sigma_{t-1|t-1}^{\ 2} A^2)} \right). \tag{5.25}$$

If we introduce

$$\sigma_{t|t-1}^{\ 2} = \sigma_v^{\ 2} + \sigma_{t-1|t-1}^{\ 2} A^2, \quad x_{t|t-1} = Ax_{t-1|t-1} + u_t, \tag{5.26}$$

We obtain

$$p(x_t \mid y_{1:t-1}) = \frac{1}{\sqrt{2\pi\sigma_{t|t-1}^{\ 2}}} \exp\left(-\frac{(x_t - x_{t|t-1})^2}{2\sigma_{t|t-1}^{\ 2}} \right). \tag{5.27}$$

The numerator $p(y_t \mid y_{1:t-1})$ of the posterior probability $p(x_t \mid y_{1:t})$ is obtained by integrating

$$p(y_t \mid y_{1:t-1}) = p(y_t \mid x_t) p(x_t \mid y_{1:t-1}) = \frac{1}{2\pi\sigma_w \sigma_{t|t-1}} \exp\left(-\frac{(y_t - Bx_t)^2}{2\sigma_w^{\ 2}} - \frac{(x_t - x_{t|t-1})^2}{2\sigma_{t|t-1}^{\ 2}} \right) \tag{5.28}$$

with respect to x_t (→Appendix 5A):

$$p(y_t \mid y_{1:t-1}) = \int_{-\infty}^{\infty} p(y_t \mid x_t) p(x_t \mid y_{1:t-1}) dx_t$$

$$= \frac{1}{2\pi\sigma_w \sigma_{t|t-1}} \int_{-\infty}^{\infty} \exp\left(-\frac{(\sigma_w^{\ 2} + \sigma_{t|t-1}^{\ 2} B^2)}{2\sigma_w^{\ 2} \sigma_{t|t-1}^{\ 2}} \left((x_t - x_{t|t-1}) - \frac{\sigma_{t|t-1}^{\ 2} B (y_t - B x_{t|t-1})}{\sigma_w^{\ 2} + \sigma_{t|t-1}^{\ 2} B^2} \right)^2 \right) dx_t \cdot$$

$$\cdot \exp\left(-\frac{(y_t - B x_{t|t-1})^2}{2(\sigma_w^{\ 2} + \sigma_{t|t-1}^{\ 2} B^2)} \right).$$

(5.29)

Since we have

$$\int_{-\infty}^{\infty} \exp\left(-\frac{(\sigma_w^{\ 2} + \sigma_{t|t-1}^{\ 2} B^2)}{2\sigma_w^{\ 2} \sigma_{t|t-1}^{\ 2}} \left((x_t - x_{t|t-1}) - \frac{\sigma_{t|t-1}^{\ 2} B (y_t - B x_{t|t-1})}{\sigma_w^{\ 2} + \sigma_{t|t-1}^{\ 2} B^2} \right)^2 \right) dx_t$$

$$= \int_{-\infty}^{\infty} \exp\left(-\frac{\sigma_w^{\ 2} + \sigma_{t|t-1}^{\ 2} B^2}{2\sigma_w^{\ 2} \sigma_{t|t-1}^{\ 2}} x_t^{\ 2} \right) dx_t = \sqrt{\frac{2\pi}{\sigma_w^{\ 2} + \sigma_{t|t-1}^{\ 2} B^2}} \sigma_w \sigma_{t|t-1} , \quad (5.30)$$

we obtain

$$p(y_t \mid y_{1:t-1}) = \frac{1}{\sqrt{2\pi(\sigma_w^{\ 2} + \sigma_{t|t-1}^{\ 2} B^2)}} \exp\left(-\frac{(y_t - B x_{t|t-1})^2}{2(\sigma_w^{\ 2} + \sigma_{t|t-1}^{\ 2} B^2)} \right). \quad (5.31)$$

If we substitute equations (5.27) and (5.31) into equation (5.17), we derive

$$p(x_t \mid y_{1:t}) = \frac{p(y_t \mid x_t) p(x_t \mid y_{1:t-1})}{p(y_t \mid y_{1:t-1})}$$

$$= \frac{1}{2\pi\sigma_w \sigma_{t|t-1}} \exp\left(-\frac{(y_t - B x_t)^2}{2\sigma_w^{\ 2}} - \frac{(x_t - x_{t|t-1})^2}{2\sigma_{t|t-1}^{\ 2}} \right) \sqrt{2\pi(\sigma_w^{\ 2} + \sigma_{t|t-1}^{\ 2} B^2)} \exp\left(\frac{(y_t - B x_{t|t-1})^2}{2(\sigma_w^{\ 2} + \sigma_{t|t-1}^{\ 2} B^2)} \right)$$

$$= \frac{\sqrt{2\pi(\sigma_w^{\ 2} + \sigma_{t|t-1}^{\ 2} B^2)}}{2\pi\sigma_w \sigma_{t|t-1}} \exp\left(-\frac{(y_t - B x_t)^2}{2\sigma_w^{\ 2}} - \frac{(x_t - x_{t|t-1})^2}{2\sigma_{t|t-1}^{\ 2}} + \frac{(y_t - B x_{t|t-1})^2}{2(\sigma_w^{\ 2} + \sigma_{t|t-1}^{\ 2} B^2)} \right).$$

(5.32)

Since we have

$$
\begin{aligned}
&-\frac{(y_t - Bx_t)^2}{2\sigma_w^2} - \frac{(x_t - x_{t|t-1})^2}{2\sigma_{t|t-1}^2} + \frac{(y_t - Bx_{t|t-1})^2}{2(\sigma_w^2 + \sigma_{t|t-1}^2 B^2)} \\
&= -\frac{(\sigma_w^2 + \sigma_{t|t-1}^2 B^2)}{2\sigma_w^2 \sigma_{t|t-1}^2} \left((x_t - x_{t|t-1}) - \frac{\sigma_{t|t-1}^2 B(y_t - Bx_{t|t-1})}{\sigma_w^2 + \sigma_{t|t-1}^2 B^2} \right)^2 - \frac{(y_t - Bx_{t|t-1})^2}{2(\sigma_w^2 + \sigma_{t|t-1}^2 B^2)} + \frac{(y_t - Bx_{t|t-1})^2}{2(\sigma_w^2 + \sigma_{t|t-1}^2 B^2)} \\
&= -\frac{(\sigma_w^2 + \sigma_{t|t-1}^2 B^2)}{2\sigma_w^2 \sigma_{t|t-1}^2} \left((x_t - x_{t|t-1}) - \frac{\sigma_{t|t-1}^2 B(y_t - Bx_{t|t-1})}{\sigma_w^2 + \sigma_{t|t-1}^2 B^2} \right)^2 ,
\end{aligned} \tag{5.33}
$$

We obtain

$$
\begin{aligned}
p(x_t \mid y_{1:t}) &= \frac{\sqrt{2\pi(\sigma_w^2 + \sigma_{t|t-1}^2 B^2)}}{2\pi\sigma_w \sigma_{t|t-1}} \exp\left(-\frac{(\sigma_w^2 + \sigma_{t|t-1}^2 B^2)}{2\sigma_w^2 \sigma_{t|t-1}^2} \left((x_t - x_{t|t-1}) - \frac{\sigma_{t|t-1}^2 B(y_t - Bx_{t|t-1})}{\sigma_w^2 + \sigma_{t|t-1}^2 B^2} \right)^2 \right) \\
&= \frac{\sqrt{2\pi(\sigma_w^2 + \sigma_{t|t-1}^2 B^2)}}{2\pi\sigma_w \sigma_{t|t-1}} \exp\left(-\frac{(\sigma_w^2 + \sigma_{t|t-1}^2 B^2)}{2\sigma_w^2 \sigma_{t|t-1}^2} \left((x_t - x_{t|t-1}) - K_t B^{-1}(y_t - Bx_{t|t-1}) \right)^2 \right) \\
&= \frac{1}{\sqrt{2\pi\sigma_{t|t}^2}} \exp\left(-\frac{(x_t - x_{t|t})^2}{2\sigma_{t|t}^2} \right).
\end{aligned} \tag{5.34}
$$

In obtaining Eq. (5.34), we introduced

$$
K_t = \frac{\sigma_{t|t-1}^2 B^2}{\sigma_w^2 + \sigma_{t|t-1}^2 B^2}, \quad x_{t|t} = x_{t|t-1} + K_t B^{-1}(y_t - Bx_{t|t-1}), \quad \sigma_{t|t}^2 = \sigma_{t|t-1}^2 B^2 - K_t \sigma_{t|t-1}^2 B^2 \tag{5.35}
$$

and use

$$
\begin{aligned}
\frac{\sigma_w^2 + \sigma_{t|t-1}^2 B^2}{2\sigma_w^2 \sigma_{t|t-1}^2 B^2} &= \frac{1}{2\sigma_{t|t-1}^2 B^2 \left(1 - \frac{\sigma_{t|t-1}^2 B^2}{\sigma_w^2 + \sigma_{t|t-1}^2 B^2} \right)} \\
&= \frac{1}{2\sigma_{t|t-1}^2 B^2 (1 - K_t)} = \frac{1}{2(\sigma_{t|t-1}^2 B^2 - K_t \sigma_{t|t-1}^2 B^2)} = \frac{1}{2\sigma_{t|t}^2}
\end{aligned} \tag{5.36}
$$

and

$$\frac{\sqrt{2\pi(\sigma_w{}^2+\sigma_{t|t-1}{}^2B^2)}}{2\pi\sigma_w\sigma_{t|t-1}}=\frac{1}{\sqrt{\pi}}\frac{\sqrt{(\sigma_w{}^2+\sigma_{t|t-1}{}^2B^2)}}{\sqrt{2}\sigma_w\sigma_{t|t-1}}$$
$$=\frac{1}{\sqrt{\pi}}\frac{1}{\sqrt{2(\sigma_{t|t-1}{}^2B^2-K_t\sigma_{t|t-1}{}^2B^2)}}=\frac{1}{\sqrt{2\pi\sigma_{t|t}{}^2}}. \tag{5.37}$$

K_t is nothing but the Kalman gain in the Kalman filter. Rewriting the mean and variance in $p(x_t \mid y_{1:t})$ given by equation (5.34), we obtain

$$x_{t|t}=\frac{\sigma_{t|t-1}{}^2B^2}{\sigma_w{}^2+\sigma_{t|t-1}{}^2B^2}B^{-1}y_t+\frac{\sigma_w{}^2}{\sigma_w{}^2+\sigma_{t|t-1}{}^2B^2}x_{t|t-1},\quad \sigma_{t|t}{}^2=\frac{\sigma_w{}^2\sigma_{t|t-1}{}^2B^2}{\sigma_w{}^2+\sigma_{t|t-1}{}^2B^2}. \tag{5.38}$$

The mean $x_{t|t}$ in the posterior distribution is a sum of the predicted value $x_{t|t-1}$ and the measured value y_t weighted by the variances $\sigma_w{}^2$ and $\sigma_{t|t-1}{}^2$. Since we have

$$\sigma_{t|t}{}^2\le\frac{\sigma_w{}^2B^2\sigma_{t|t-1}{}^2}{\sigma_{t|t-1}{}^2B^2}=\sigma_w{}^2,\quad \sigma_{t|t}{}^2\le\frac{\sigma_w{}^2B^2\sigma_{t|t-1}{}^2}{\sigma_w{}^2}=B^2\sigma_{t|t-1}{}^2, \tag{5.39}$$

the variance $\sigma_{t|t}{}^2$ of the posterior distribution is smaller than the variance $B^2\sigma_{t|t-1}{}^2$ predicted from the variance $\sigma_{t|t-1}{}^2$ of the prior distribution.

We summarize the above results:

(1) One step ahead prediction

$$x_{t|t-1}=Ax_{t-1|t-1}+u_t, \tag{5.40}$$
$$\sigma_{t|t-1}{}^2=\sigma_v{}^2+\sigma_{t-1|t-1}{}^2A^2. \tag{5.41}$$

(2) Filtering

$$x_{t|t}=x_{t|t-1}+K_tB^{-1}(y_t-Bx_{t|t-1}), \tag{5.42}$$
$$\sigma_{t|t}{}^2=\sigma_{t|t-1}{}^2B^2-K_t\sigma_{t|t-1}{}^2B^2. \tag{5.43}$$

(3) Karman gain

$$K_t=\frac{\sigma_{t|t-1}{}^2B^2}{\sigma_w{}^2+\sigma_{t|t-1}{}^2B^2}. \tag{5.44}$$

5.3.2. Results of numerical calculation

Using the theory of sequential estimation in section 5.3.1, the numerical calculation results by the maximum likelihood method (a special case of Bayesian inference) described in section 5.1.2 (b) are shown below. That is, we define a function $P(\hat{A},\hat{B})$ is defined from the observation model in equation (5.2):

$$P(\hat{A},\hat{B}) = \prod_{t=1}^{T} \frac{1}{\sqrt{2\pi\sigma_w{}^2}} \exp\left(-\frac{(y_t - \hat{B}x_{t|t})^2}{2\sigma_w{}^2} \right). \tag{5.45}$$

Taking the natural logarithm $\log P(\hat{A},\hat{B})$:

$$\log P(A,B) = -\frac{T}{2}\log\left(2\pi\sigma_w{}^2\right) - \sum_{t=1}^{T} \frac{(y_t - Bx_{t|t})^2}{2\sigma_w{}^2}, \tag{5.46}$$

we obtain $\hat{A}$ and $\hat{B}$ that maximize $\log P(A,B)$ and make them as estimates of $\hat{A}$ and $\hat{B}$. We apply the steepest ascent method for the search of the maximum. The differentiation is conducted numerically:

$$\begin{aligned} \frac{\log P(\hat{A},\hat{B})}{\partial A} &= \lim_{d\lambda \to 0} \frac{\log P(\hat{A}+d\lambda_A,\hat{B}) - \log P(\hat{A},\hat{B})}{d\lambda_A}, \\ \frac{\log P(\hat{A},\hat{B})}{\partial B} &= \lim_{d\lambda \to 0} \frac{\log P(\hat{A},\hat{B}+d\lambda_B) - \log P(\hat{A},\hat{B})}{d\lambda_B}. \end{aligned} \tag{5.47}$$

We used the following values for the parameters. Set values for generation of the observation data:

$$A = B = 1,\ \sigma_V = 0.0,\ \sigma_W = 0.05, \tag{5.48a}$$

Set values for convergence calculation:

$$\hat{A} = 0.3,\ \hat{B} = 0.5,\ d\lambda_A = d\lambda_B = 1\times10^{-7},\ dl = 1\times10^{-7}. \tag{5.48b}$$

The numerical results are shown in Fig. 5 and Fig. 6. In the figures, not only the unknown parameters A and B but also the state x converge to the set value by the data assimilation. The convergence seems good.

The programming code is shown in Appendix 5C.

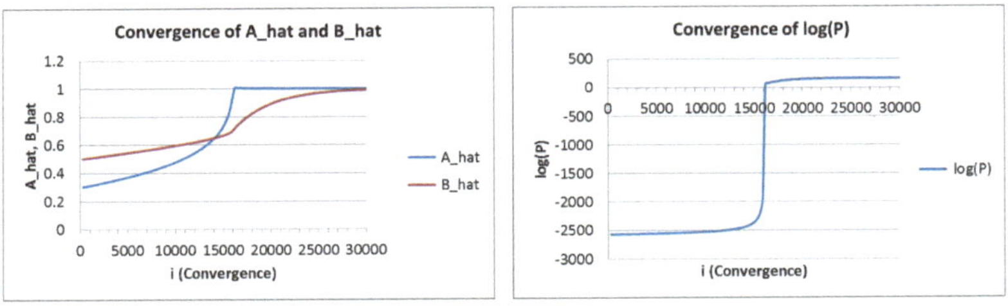

Fig. 5.5. Convergence of Mountain Climb Method in Search of Solution (SSM+Bayes).

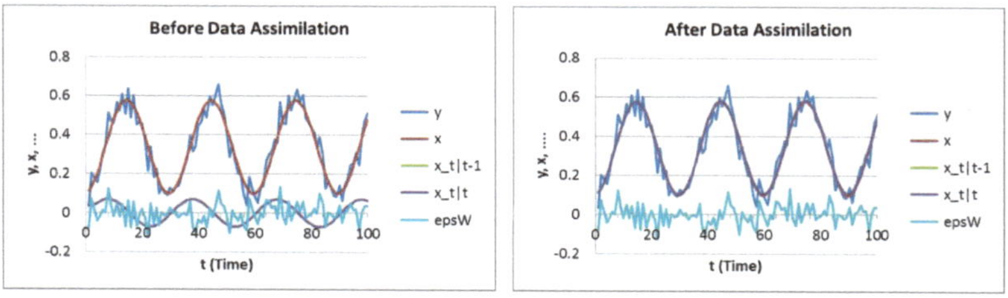

Fig. 5.6. Solution by Data Assimiltion (SSM+Bayes).

Numerical simulation technology has made remarkable progress. Calculation accuracy, speed, and scale of the target have improved dramatically. Complex fluid motion, combustion problems so on can be calculated using state-of-the-art supercomputers.

On the other hand, the limits of numerical simulation have become clear. If the object is not captured in a completely explicit form, the problem could not be dealt with satisfactorily. Even in such a case, the situation will be different if there is something to supplement the unknown part, for example, observation data. By combining observation data with numerical simulation, it would be possible to grasp the phenomenon with considerable accuracy.

The effectiveness of combining data observation and numerical simulation was recognized, and it came to be called data assimilation technology, and its application in various fields became active. However, understanding data assimilation requires knowledge of phenomena and their mathematics and the underlying statistics of assimilation, especially Bayesian statistics. Especially the part related to Bayesian statistics is advanced and not easy to understand.

The part about Bayesian statistics requires a high degree of knowledge because there is a convenient but not easy time-series sequential estimation using Bayesian estimation. Sequential estimation by Bayesian estimation is not essential for understanding the concept of data assimilation itself.

Data assimilation consists of two parts: a time series calculation part and an assimilation part that estimates unknown parameters. In this paper, we first explain how to perform sequential estimation of time series directly with a system model, ignoring system noise, and perform data assimilation by the least square method or maximum likelihood method. We hope that the explanations in this book will help beginners understand.

REFRERENCES 5

[1] Anderson, B. D. and Moore, J. B. (1979). Optimal Filtering, Prentice-Hall, Englewood Cliffs

[2] Kitagawa, G. (1984). State space modeling of nonstationary time series and smoothing of unequally spaced data, Time Series Analysis of Irregularly Observed Data, 189-210, Springer, New York.

[3] Fukaya K, Royle JA (2013) Markov models for community dynamics allowing for observation error. Ecology, 94:2670- 2677
[4] Kody Law, Andrew Stuart and Konstantinos Zygalakis: "Data Assimilation - A Mathematical Introduction", Springer Int. Pub., ISBN 978-3-319-20325-6 (2015).

Appendix 5A Derivation of Probability Density Distribution $p(x_t \mid y_{1:t-1})$ given by equation (5.23)

$$
\begin{aligned}
p(x_t \mid y_{1:t-1}) &= \frac{1}{2\pi\sigma_v\sigma_{t-1|t-1}}\int_{-\infty}^{\infty}\exp\left(-\frac{(x_t - Ax_{t-1} - u_t)^2}{2\sigma_v^2} - \frac{(x_{t-1} - x_{t-1|t-1})^2}{2\sigma_{t-1|t-1}^2}\right)dx_{t-1} \\
&= \frac{1}{2\pi\sigma_v\sigma_{t-1|t-1}}\int_{-\infty}^{\infty}\exp\left(-\frac{\sigma_{t-1|t-1}^2\left(x_t - Ax_{t-1|t-1} - u_t - A(x_{t-1} - x_{t-1|t-1})\right)^2 + \sigma_v^2(x_{t-1} - x_{t-1|t-1})^2}{2\sigma_v^2\sigma_{t-1|t-1}^2}\right)dx_{t-1} \\
&= \frac{1}{2\pi\sigma_v\sigma_{t-1|t-1}}\int_{-\infty}^{\infty}\exp\left(-\frac{1}{2\sigma_v^2\sigma_{t-1|t-1}^2}\left\{\begin{aligned}&\sigma_{t-1|t-1}^2\begin{bmatrix}(x_t - Ax_{t-1|t-1} - u_t)^2 \\ -2A(x_t - Ax_{t-1|t-1} - u_t)(x_{t-1} - x_{t-1|t-1}) \\ +A^2(x_{t-1} - x_{t-1|t-1})^2\end{bmatrix} \\ &+\sigma_v^2(x_{t-1} - x_{t-1|t-1})^2\end{aligned}\right\}\right)dx_{t-1} \\
&= \frac{1}{2\pi\sigma_v\sigma_{t-1|t-1}}\int_{-\infty}^{\infty}\exp\left(-\frac{1}{2\sigma_v^2\sigma_{t-1|t-1}^2}\left\{\begin{aligned}&(\sigma_v^2 + \sigma_{t-1|t-1}^2A^2)(x_{t-1} - x_{t-1|t-1})^2 \\ &-2\sigma_{t-1|t-1}^2A(x_t - Ax_{t-1|t-1} - u_t)(x_{t-1} - x_{t-1|t-1}) \\ &+\sigma_{t-1|t-1}^2(x_t - Ax_{t-1|t-1} - u_t)^2\end{aligned}\right\}\right)dx_{t-1} \\
&= \frac{1}{2\pi\sigma_v\sigma_{t-1|t-1}}\int_{-\infty}^{\infty}\exp\left(-\frac{1}{2\sigma_v^2\sigma_{t-1|t-1}^2}\left\{\begin{aligned}&(\sigma_v^2 + \sigma_{t-1|t-1}^2A^2)\begin{bmatrix}(x_{t-1} - x_{t-1|t-1}) \\ -\left(\dfrac{\sigma_{t-1|t-1}^2A(x_t - Ax_{t-1|t-1} - u_t)}{\sigma_v^2 + \sigma_{t-1|t-1}^2A^2}\right)\end{bmatrix}^2 \\ &-(\sigma_v^2 + \sigma_{t-1|t-1}^2A^2)\left(\frac{\sigma_{t-1|t-1}^2A(x_t - Ax_{t-1|t-1} - u_t)}{\sigma_v^2 + \sigma_{t-1|t-1}^2A^2}\right)^2 \\ &+\sigma_{t-1|t-1}^2(x_t - Ax_{t-1|t-1} - u_t)^2\end{aligned}\right\}\right)dx_{t-1} \\
&= \frac{1}{2\pi\sigma_v\sigma_{t-1|t-1}}\int_{-\infty}^{\infty}\exp\left(-\frac{\sigma_v^2 + \sigma_{t-1|t-1}^2A^2}{2\sigma_v^2\sigma_{t-1|t-1}^2}\left[(x_{t-1} - x_{t-1|t-1}) - \left(\frac{\sigma_{t-1|t-1}^2A(x_t - Ax_{t-1|t-1} - u_t)}{\sigma_v^2 + \sigma_{t-1|t-1}^2A^2}\right)\right]^2\right)dx_{t-1} \cdot \\
&\quad \cdot\exp\left(-\frac{(x_t - Ax_{t-1|t-1} - u_t)^2}{2(\sigma_v^2 + \sigma_{t-1|t-1}^2A^2)}\right)
\end{aligned}
$$

Appendix 5B Derivation of Probability Density Distribution $p(y_t \mid y_{1:t-1})$ given by equation (5.29)

$$p(y_t \mid y_{1:t-1}) = \int_{-\infty}^{\infty} p(y_t \mid x_t) p(x_t \mid y_{1:t-1}) dx_t$$

$$= \frac{1}{2\pi\sigma_w \sigma_{t|t-1}} \int_{-\infty}^{\infty} \exp\left(-\frac{(y_t - Bx_t)^2}{2\sigma_w^2} - \frac{(x_t - x_{t|t-1})^2}{2\sigma_{t|t-1}^2} \right) dx_t$$

$$= \frac{1}{2\pi\sigma_w \sigma_{t|t-1}} \int_{-\infty}^{\infty} \exp\left(\frac{1}{2\sigma_w^2 \sigma_{t|t-1}^2} \left(-\sigma_{t|t-1}^2 \left(y_t - Bx_{t|t-1} - B(x_t - x_{t|t-1}) \right)^2 - \sigma_w^2 (x_t - x_{t|t-1})^2 \right) \right) dx_t$$

$$= \frac{1}{2\pi\sigma_w \sigma_{t|t-1}} \int_{-\infty}^{\infty} \exp\left(\frac{1}{2\sigma_w^2 \sigma_{t|t-1}^2} \begin{pmatrix} -(\sigma_w^2 + \sigma_{t|t-1}^2 B^2)(x_t - x_{t|t-1})^2 \\ +2\sigma_{t|t-1}^2 B(y_t - Bx_{t|t-1})(x_t - x_{t|t-1}) \\ -\sigma_{t|t-1}^2 (y_t - Bx_{t|t-1})^2 \end{pmatrix} \right) dx_t$$

$$= \frac{1}{2\pi\sigma_w \sigma_{t|t-1}} \int_{-\infty}^{\infty} \exp\left(\frac{1}{2\sigma_w^2 \sigma_{t|t-1}^2} \begin{pmatrix} -(\sigma_w^2 + \sigma_{t|t-1}^2 B^2) \left((x_t - x_{t|t-1}) - \frac{\sigma_{t|t-1}^2 B(y_t - Bx_{t|t-1})}{\sigma_w^2 + \sigma_{t|t-1}^2 B^2} \right)^2 \\ + \frac{\sigma_{t|t-1}^4 B^2 (y_t - Bx_{t|t-1})^2}{\sigma_w^2 + \sigma_{t|t-1}^2 B^2} - \sigma_{t|t-1}^2 (y_t - Bx_{t|t-1})^2 \end{pmatrix} \right) dx_t$$

$$= \frac{1}{2\pi\sigma_w \sigma_{t|t-1}} \int_{-\infty}^{\infty} \exp\left(-\frac{(\sigma_w^2 + \sigma_{t|t-1}^2 B^2)}{2\sigma_w^2 \sigma_{t|t-1}^2} \left((x_t - x_{t|t-1}) - \frac{\sigma_{t|t-1}^2 B(y_t - Bx_{t|t-1})}{\sigma_w^2 + \sigma_{t|t-1}^2 B^2} \right)^2 \right) dx_t \cdot$$

$$\cdot \exp\left(-\frac{(y_t - Bx_{t|t-1})^2}{2(\sigma_w^2 + \sigma_{t|t-1}^2 B^2)} \right)$$

Appendix 5C Code for 1d state space model

"Microsoft C/C++ Compiler Version 17.00.50727.1 for x86" and "Microsoft Linker Version 11.00.50727" were used for compile and link (in command window; cl source_file_name.c).

(1) Programing code: _1dStateSpaceModelBayes.c

```
// ------------------------------------------------------------------ //
//                                                                    //
// File Name: _1dStateSpaceModel.c               2021.04.29-2021.05.06 //
// File Name: _1dStateSpaceModelOpt.c            2021.05.06-2021.05.07 //
// File Name: _1dStateSpaceModelBayes.c          2021.05.06-2021.05.11 //
// File Name: _1dStateSpaceModelBayes.c          2022.05.24-2022.05.24 //
//                                                                    //
//    1D State Space Model                                            //
//                                                                    //
// ------------------------------------------------------------------ //

// ----- functions --------------------------------------------------- //

#include <stdio.h>
#include <stdlib.h>
#include <string.h>
#include <math.h>

#define PI     3.14159265    // Pi

void main();
void pushKey();

double Uniform( void );                      // uniform random number [0,1]
double rand_normal(double,double);           // normal random number

double normal(double,double,double);         // normal distribution

double Pnml_I(int n, double c, double myu, double sgm);       //
double logPnml_I(int n, double c, double myu, double sgm);    //

void BayesPrediction(double A, double B, double sgmV, double sgmW, int T,
                     double *y, double *x, double *x_tBtm1, double *x_tBt,
```

```
                    double *epsW, double *sgm_tBtm1, double *sgm_tBt, double *K);

// ----- variables ----------------------------------------------- //

char title_memo[5000];

FILE *fp_inp;                         // file pointer of input file
FILE *fp_out;                         // file pointer of output file

char InputDataFile[80];               // input file name
char OutputDataFile[80];              // output file name

char buf[5000];

int T;
int t;

double A1;
double B1;
double A;
double B;
double u[1001];
double x[1001];
double x_tBtm1[1001];
double x_tBt[1001];
double y[1001];
double epsW[1001];
double sgm_tBtm1[1001];
double sgm_tBt[1001];
double sgm2_tBtm1[1001];
double sgm2_tBt[1001];
double sgmV;
double sgmW;
double K[1001];

double A_;
double B_;

double lmd;                           // step of mountain climbing
```

```
double dlmdA;                   // difference in derivative
double dlmdB;                   // difference in derivative
int iEnd=100;

int iSkp;                       // skip print lines
int nSkp;                       // skip print lines for process results
int oSkp;                       // skip observation

double logRProb_;               // log of RProb_
double logRProb_A;              // derivative of logRProb_ wrt A
double logRProb_B;              // derivative of logRProb_ wrt B

int Acon;                       // control infectant
int Bcon;                       // control infectant

int PrtCtrl;                    // print control ... 1: print


// -------------------------------------------------------------- //

void main()
{

    int i, j, k;

    //// open input file
    sprintf(InputDataFile, "_1dStateSpaceModelBayes_inp.dat");

    if ((fp_inp = fopen(InputDataFile, "r")) == NULL) {
        printf("Failed in Reading Input Data File! ... %s¥n", InputDataFile);
        exit(1);
    }

    //// open output file
   sprintf(OutputDataFile, "_1dStateSpaceModelBayes_out.csv");

    if ((fp_out = fopen(OutputDataFile, "w")) == NULL) {
        printf("Failed in Reading Output Data File! ... %s¥n", OutputDataFile);
        exit(1);
```

```
}

//// input from file
fscanf(fp_inp, "%s", title_memo);

fscanf(fp_inp, "%s %d", buf, &T);

fscanf(fp_inp, "%s %lf", buf, &A1);
fscanf(fp_inp, "%s %lf", buf, &B1);

fscanf(fp_inp, "%s %lf", buf, &A);
fscanf(fp_inp, "%s %lf", buf, &B);

fscanf(fp_inp, "%s %lf", buf, &sgmV);
fscanf(fp_inp, "%s %lf", buf, &sgmW);

fscanf(fp_inp, "%s %d", buf, &Acon);
fscanf(fp_inp, "%s %d", buf, &Bcon);

fscanf(fp_inp, "%s %lf", buf, &dlmdA);
fscanf(fp_inp, "%s %lf", buf, &dlmdB);
fscanf(fp_inp, "%s %lf", buf, &lmd);

fscanf(fp_inp, "%s %d", buf, &iEnd);
fscanf(fp_inp, "%s %d", buf, &iSkp);
fscanf(fp_inp, "%s %d", buf, &nSkp);
fscanf(fp_inp, "%s %d", buf, &oSkp);
fscanf(fp_inp, "%s %d", buf, &PrtCtrl);

fclose(fp_inp);


printf("memo: %s¥n", title_memo);
printf("¥n");

printf("T       = %d¥n", T);

printf("A1      = %12.6f¥n", A1);
printf("B1      = %12.6f¥n", B1);
```

```
printf("A       = %12.6f¥n", A);
printf("B       = %12.6f¥n", B);

printf("sgmV    = %12.6f¥n", sgmV);
printf("sgmW    = %12.6f¥n", sgmW);

printf("Acon         = %d¥n", Acon);
printf("Bcon         = %d¥n", Bcon);

printf("dlmdA        = %12.9f¥n", dlmdA);
printf("dlmdB        = %12.9f¥n", dlmdB);
printf("lmd          = %12.9f¥n", lmd);

printf("iEnd         = %d¥n", iEnd);
printf("iSkp         = %d¥n", iSkp);
printf("nSkp         = %d¥n", nSkp);
printf("oSkp         = %d¥n", oSkp);
printf("PrtCtrl      = %d¥n", PrtCtrl);
printf("¥n");


fprintf(fp_out, "memo: %s¥n", title_memo);
fprintf(fp_out, "¥n");

fprintf(fp_out, "T =, %d¥n", T);

fprintf(fp_out, "A1 =, %12.6f¥n", A1);
fprintf(fp_out, "B1 =, %12.6f¥n", B1);

fprintf(fp_out, "A =, %12.6f¥n", A);
fprintf(fp_out, "B =, %12.6f¥n", B);

fprintf(fp_out, "sgmV =, %12.6f¥n", sgmV);
fprintf(fp_out, "sgmW =, %12.6f¥n", sgmW);

fprintf(fp_out, "Acon =, %d¥n", Acon);
fprintf(fp_out, "Bcon =, %d¥n", Bcon);
```

```
    fprintf(fp_out, "dlmdA =, %12.9f¥n", dlmdA);
    fprintf(fp_out, "dlmdB =, %12.9f¥n", dlmdB);
    fprintf(fp_out, "lmd =, %12.9f¥n", lmd);

    fprintf(fp_out, "iEnd =, %d¥n", iEnd);
    fprintf(fp_out, "iSkp =, %d¥n", iSkp);
    fprintf(fp_out, "nSkp =, %d¥n", nSkp);
    fprintf(fp_out, "oSkp =, %d¥n", oSkp);
    fprintf(fp_out, "PrtCtrl =, %d¥n", PrtCtrl);
    fprintf(fp_out, "¥n");


    for (t = 1; t <= T; t++)
        u[t] = 0.05*sin(2.0*3.1416/30.0*t);

    // data generation

    x[0] = 0.1;
    for (t = 1; t <= T; t++) {
        x[t] = A1*x[t-1]+u[t];
        epsW[t] = rand_normal(0.0, sgmW);
        y[t] = B1*x[t] + epsW[t]; ////////////////////
    }


    x_tBt[0] = 0.1;
    sgm2_tBtm1[0] = 0.0;

    BayesPrediction(A, B, sgmV, sgmW, T, y, x, x_tBtm1, x_tBt, epsW, sgm_tBtm1, sgm_tBt, K);

    fprintf(fp_out, "t, y, x, x_tBtm1, x_tBt, epsW, sgm_tBtm1, sgm_tBt, K¥n");
    for (t = 1; t <= T; t++)
        fprintf(fp_out,
"%d, %12.6f, %12.6f, %12.6f, %12.6f, %12.6f, %12.6f, %12.6f, %12.6f¥n",
                t, y[t], x[t], x_tBtm1[t], x_tBt[t], epsW[t], sgm_tBtm1[t], sgm_tBt[t], K[t]);
    fprintf(fp_out, "¥n");

    A_ = A;
    B_ = B;
```

```
    fprintf(fp_out, "i, A_, B_, logRProb, logRProb_A, logRProb_B¥n");
    for (i = 1; i <= iEnd; i++) {


        BayesPrediction(A_, B_, sgmV, sgmW, T, y, x, x_tBtm1, x_tBt, epsW, sgm_tBtm1, sgm_tBt,
K);
        logRProb_ = 0.0;
        for (t = 1; t <= T; t++) {
            logRProb_ += logPnml_I(t, B_, 0.0, sgmW);
        }

        if (Acon == 1) {
            BayesPrediction(A_+dlmdA, B_, sgmV, sgmW, T, y, x, x_tBtm1, x_tBt, epsW, sgm_tBtm1,
sgm_tBt, K);
            logRProb_A = 0.0;
            for (t = 1; t <= T; t++)
                logRProb_A += logPnml_I(t, B_, 0.0, sgmW);
            logRProb_A = (logRProb_A-logRProb_)/dlmdA;
        }

        if (Bcon == 1) {
            BayesPrediction(A_, B_+dlmdB, sgmV, sgmW, T, y, x, x_tBtm1, x_tBt,
                            epsW, sgm_tBtm1, sgm_tBt, K);
            logRProb_B = 0.0;
            for (t = 1; t <= T; t++)
                logRProb_B += logPnml_I(t, B_+dlmdB, 0.0, sgmW);
            logRProb_B = (logRProb_B-logRProb_)/dlmdB;
        }

        if (Acon == 1)
            A_ += logRProb_A*lmd;
        if (Bcon == 1)
            B_ += logRProb_B*lmd;

        if (i % iSkp == 0)
            fprintf(fp_out, "%d, %12.6f, %12.6f, %12.6f, %12.6f, %12.6f¥n",
                    i, A_, B_, logRProb_, logRProb_A, logRProb_B);
    }
    fprintf(fp_out, "¥n");
```

```
    BayesPrediction(A_, B_, sgmV, sgmW, T, y, x, x_tBtm1, x_tBt, epsW, sgm_tBtm1, sgm_tBt, K);

    fprintf(fp_out, "t, y, x, x_tBtm1, x_tBt, epsW, sgm_tBtm1, sgm_tBt, K¥n");
    for (t = 1; t <= T; t++)
        fprintf(fp_out,
"%d, %12.6f, %12.6f, %12.6f, %12.6f, %12.6f, %12.6f, %12.6f, %12.6f¥n",
                t, y[t], x[t], x_tBtm1[t], x_tBt[t], epsW[t], sgm_tBtm1[t], sgm_tBt[t], K[t]);
    fprintf(fp_out, "¥n");


    fclose(fp_out);

    pushKey();

}

// ------------------------------------------------------------------ //

void pushKey()
{
    printf("¥n      Push Return Key! ");
    getchar();
    getchar();
}

// ------------------------------------------------------------------ //

double rand_normal( double myu, double sigma )
{
    double z=sqrt( -2.0*log(Uniform()) ) * sin( 2.0*PI*Uniform() );
    return myu + sigma*z;
 }

// ------------------------------------------------------------------ //

double Uniform( void )
{
    static int x=10;
    int a=1103515245,b=12345,c=2147483647;
```

```
    x = (a*x + b)&c;

    return ((double)x+1.0) / ((double)c+2.0);
}

// ------------------------------------------------------------------ //

double normal(double x, double myu, double sgm)      // normal distribution
{
    return 1.0/sqrt(2.0*PI*sgm*sgm)*exp(-(x-myu)*(x-myu)/2.0/sgm/sgm);
}

// ------------------------------------------------------------------ //

double Pnml_I(int t, double B, double myu, double sgm)
{
    return normal(y[t]-B*x_tBt[t], myu, sgm);
}

// ------------------------------------------------------------------ //

double logPnml_I(int t, double B, double myu, double sgm)
{
    return log(Pnml_I(t, B, myu, sgm));
}

// ------------------------------------------------------------------ //

void BayesPrediction(double A, double B, double sgmV, double sgmW, int T,
                    double *y, double *x, double *x_tBtm1, double *x_tBt,
                    double *epsW, double *sgm_tBtm1, double *sgm_tBt, double *K) {
    int t;

    for (t = 1; t <= T; t++) {
        x_tBtm1[t] = A*x_tBt[t-1]+u[t];
        sgm2_tBtm1[t] = sgmV*sgmV + sgm2_tBt[t-1]*A*A;    /*****/

        K[t] = sgm2_tBtm1[t]*B*B/(sgmW*sgmW+sgm2_tBtm1[t]*B*B);
```

```
        x_tBt[t] = x_tBtm1[t] + K[t]/B*(y[t]-B*x_tBtm1[t]);
        sgm2_tBt[t] = sgm2_tBtm1[t]*B*B*(1.0 - K[t]);

        sgm_tBtm1[t] = sqrt(sgm2_tBtm1[t]);
        sgm_tBt[t] = sqrt(sgm2_tBt[t]);
    }
}

// ---------------------------------------------------------------- //
```

(2) Input file: _1dStateSpaceModelBayes_inp.dat

```
_1dStateSpaceModel_20210503
T                       100
A1                      1.0
B1                      1.0
A                       0.3
B                       0.5
sgmV                    0.0
sgmW0.01                0.05
Acon                    1
Bcon                    1
dlmdA                   0.0000001
dlmdB                   0.0000001
lmd                     0.0000001
iEnd30000               30000
iSkp100                 300
nSkp                    10
oSkp:observationSkip    10
PrtCtrl                 0
```

6. SOLUTION OF SIR INFECTION EQUATION USING DATA ASSIMILATION

The new coronavirus infection (COVID-19) is rampant. Even now, more than a year after the start of the infection, the infection has not yet shown signs of ending. It can be said to be an infection that remains in history.

The most troublesome part of this infection is that it not only takes several days from infection to onset, but it also infects for several days even in the not-onset state. Therefore, a considerable number of infected persons with infectivity are left unchecked. Therefore, even if the infection status is simulated by the SIR equation [1-2], the true values of the infection parameters and the number of infected persons cannot be grasped.

However, it is possible to observe the infection status as it is. The daily number of infected people and the cumulative number of infected people are announced. The numbers in these data are not true values, but they reflect true values. It is very useful for getting a rough idea of the infection status.

As mentioned above, it is impossible to grasp the true value only by the SIR equation [1-2]. However, it may be possible to estimate the true value by combining it with the observation equation. In short, the framework of data assimilation or state-space model [3-5] is considered to be effective. The parameters of the SIR equation, such as the daily number of infected people, infection rate, and recovery rate, are unknown but are embedded in the observed values. Unknown parameters must consist with the observed values. Wouldn't it be possible to determine these unknowns by combining them using Bayesian inference or the method of least squares? We report this effectiveness because we were able to confirm this effectiveness from the numerical results.

In order to fully explain the infection phenomenon, it is necessary to explain spatial characteristics such as the effect of the population density distribution, but this is not covered here. We focus on elucidating recurring infection waves and hidden infections. Regarding the spatial characteristics of the SIR equation, we would appreciate it if you could refer to the author's paper [2] in References.

6.1. State space model of SIR infection equation

6.1.1. State Space Model

Among the physical models we are targeting, there are many that we know the differential equations (system models) describing the physical phenomena (systems) but cannot directly observe the physical phenomena themselves or observe them sufficiently. An earthquake-like phenomenon would be a good example. It is not fully understood what kind of process is going on underground. There is a weather forecast around us. The reliable forecasts might be difficult with theory alone.

However, even in such a case, there might be observation data reflecting a physical phenomenon. Although it might be insufficient, it might help to solve the problem. A state-space model that makes predictions using both theory and observational data gives us an idea of such cases. Originally born in control engineering, it is now considered to be a means of improving the accuracy of numerical simulations, and has come to be widely used in many fields other than control as data assimilation. The best known would be the weather forecast.

Let the system variable $x(t)$ be a function of time t, and let the observed variable be $y(t)$. For simplicity, we consider one-dimensional case. In the state space model, the system model that describes the physical system is given by

$$\frac{dx}{dt} = ax + bu + \varepsilon_x, \tag{6.1}$$

and the observation model by

$$y = cx + du + \varepsilon_y, \tag{6.2}$$

where $u(t)$ is an external input, a, b, c, d are parameters, and $\varepsilon_x(t)$ and $\varepsilon_y(t)$ are noises following normal distributions with mean 0 and variances σ_x^2 and σ_y^2. Namely, in one-dimensional case, we can assume

$$\varepsilon_x \sim \frac{1}{\sqrt{2\pi\sigma_x^2}} \exp\left(-\frac{\varepsilon_x^2}{2\sigma_x^2}\right), \quad \varepsilon_y \sim \frac{1}{\sqrt{2\pi\sigma_y^2}} \exp\left(-\frac{\varepsilon_y^2}{2\sigma_y^2}\right). \tag{6.3}$$

The non-linear case may be considered, but the linear case may be sufficient for the essential discussion. However, the SIR infection equation actually discussed below is multidimensional and non-linear. When the noise ε_x does not exist or can be negligible, we have

$$\frac{dx}{dt} = ax + bu, \tag{6.4}$$

$$y = cx + du + \varepsilon_y. \tag{6.5}$$

Only the observed noise ε_y exists as noise. In this case, since it is easy to handle, a mathematical model is made using Eqs. (6.4) and (6.5). The application to the SIR infection equation described below is also considered in this direction.

In this paper, we use the likelihood method (a special case of Bayesian inference) for data assimilation to solve the above-mentioned state-space model.

6.1.2. State space model of SIR equation

The infection phenomenon caused by the new coronavirus infection COVID-19 follows the mean-field theory called the SIR equation:

$$\frac{dS}{dt} = -\frac{\beta}{N} SI, \tag{6.6}$$

$$\frac{dI}{dt} = \frac{\beta}{N} SI - \gamma I, \tag{6.7}$$

$$\frac{dR}{dt} = \gamma I. \tag{6.8}$$

N, S, I, and *R* are the total population, the uninfected population (strictly speaking, the susceptible population), the currently infected population, and the recovered population (including the dead). From equations (6), (7), and (8), the following population conservation equations are obtained:

$$N = S + I + R, \tag{6.9}$$

β/N and γ are infection and recovery rates, respectively. The first term on the right-hand side of equations (6) and (7) is generally given in the form of not divided by *N*, but as will be described later, in order to reduce the population dependence of β, it should be considered in the form of divided by *N*.

If persons found infected are quarantined, the infectivity can be contained. Hence, that amount must be subtracted from the presently infected persons *I* in the calculation of newly infected persons. Applying SIR theory in such cases seems problematic. However, the SIR theory could be considered sufficient for the purpose of theoretically examining the possibility of data assimilation.

The following four points give difficulties to handle the infection phenomenon by the SIR equation:

(1) Since it is a mean-field theory, it is not possible to express the influence of spatial distribution characteristics such as population density.

(2) Since there are unidentified infected persons, the true number of infected persons is unknown.

(3) Since there are waves in the infection phenomenon, it is necessary to introduce the mechanism of the waves.

(4) The effect of people coming and going cannot be clearly expressed.

On the other hand, daily time-series data on newly infected persons, dead persons, severely ill persons, etc. are published daily. New daily infections are not true daily infections, as some of them are unidentified. Severely ill people might be fairly close to the true value, and dead people would be the almost true value.

Regarding (1), it is sufficient to extend the theory of spatial discretization such as dividing a large population group into smaller population groups. Regarding (2), it is impossible to observe the true number of the infected person, but since there is observation data that reflects it, it is conceivable to use this. Namely, the infection phenomenon will be modeled within the framework of the state space model. In the following, we will consider from this point of view.

Regarding (3), when the number of infections decreases, factors that promote infection such as fewer people wearing masks are created due to the relaxation of people, and when the number of infections increases, factors that suppress infection are created. Regarding (4), the more people contact, the more infections, and the fewer people contact the fewer infections. Therefore, since infections increase or decrease due to social influences, the infection coefficient should be regarded as an effective coefficient that reflects these effects.

If x in Eq. (6.4) is considered as a three-dimensional vector and extended to the non-linear case, it can be used as a system equation for a state-space model of the present case.

On the other hand, observational data include daily newly infected person O_{DlyI}, severely ill person O_{SrsI}, and dead persons O_{Dead}. The currently infected person $O_{PrsI}(\sim I)$ and the cumulative infected person O_{AccI} could also be used as observation data. Of course, it would be possible to use multiple observation data, but in this paper, we will limit it to one observation data. For simplicity, the observation model shall use only the cumulative number of infected individuals for example:

$$O_{AccI\,n} = cI_{Acc\,n} + \varepsilon_n, \quad (n = 1, 2, \cdots, N). \tag{6.10}$$

n is the time step, and time t is $t = n\Delta t \; (\Delta t = 1\text{day})$.

Even if only equation (6.10) is used as the observation model, the effects of unknown variables such as the number of initial infections $I(0)$, infection rate β, recovery rate γ, and variance σ^2 are reflected in the observation data. Furthermore, very importantly, the observation coefficient c can also be an unknown variable in this model.

6.1.3. Solution of SIR state space model using Bayes inference

If the observed data is O and the unknown is $U_j, (j = 1, 2, \cdots, J)$, the conditional probability is given by Bayes' theorem:

$$P(U_j \mid O) = \frac{P(O, U_j)}{P(O)} = \frac{P(O, U_j)}{\sum_j P(O, U_j)} = \frac{P(O \mid U_j) P(U_j)}{\sum_j P(O \mid U_j) P(U_j)}. \tag{6.11}$$

$P(U_j)$, $P(O \mid U_j)$, and $P(U_j \mid O)$ are called prior probabilities, likelihood functions, and posterior probabilities, respectively. This makes it possible to calculate the probability of an unknown quantity that yields observational data. Bayesian inference infers that U_j, which maximizes this posterior probability, gives O.

If it is allowed to assume

$$P(U_1) = P(U_2) = \cdots = P(U_J), \tag{6.12}$$

Equation (6.11) becomes

$$P(U_j \mid O) = \frac{P(O,U_j)}{\sum_j P(O,U_j)}. \tag{6.13}$$

This is nothing but the likelihood estimation. $p(U_j \mid O)$ becomes maximum when $p(U_j \mid O)$ is maximum. In the following, this assumption is used.

In the calculation of likelihood function $P(O \mid U_j)$, in case of $O = O_{Acc\,I}$ for example, equation (6.10) is used. Namely

$$P\big((O_{AccI\,n} - cI_{AccI\,n})/(cI_{AccI\,n})\big) = N(0,\sigma^2) \quad \text{or} \quad P\big(O_{AccI\,n}/(cI_{AccI\,n})\big) = N(1,\sigma^2). \tag{6.14}$$

The effects of the number of initial infections $I(0)$, infection rate β and recovery rate γ are reflected in the accumulated number of infections $I_{AccI\,n}, (n = 1,2,\cdots,N)$. Using equation (6.14), we have

$$P(O \mid U_j) = \prod_{n=1}^{N} \frac{1}{\sqrt{2\pi\sigma^2}} \exp\left(-\frac{(O_{AccI\,n} - cI_{AccI\,n})^2/(cI_{AccI\,n})^2}{2\sigma^2} \right). \tag{6.15}$$

The maximum likelihood method is used for the estimation of unknown parameters. That is, the one that maximizes the likelihood function is found by the steepest ascent method. However, as can be seen from equation (6.15), if $P(O \mid U_j)$ itself is used, there is a high possibility that underflow will occur, so we consider taking the natural logarithm $\log P(O \mid U_j)$. Namely, the maximum of

$$\log P(O \mid U_j) = N \log\left(\frac{1}{\sqrt{2\pi\sigma^2}} \right) - \sum_{n=1}^{N} \frac{(O_{AccI\,n} - cI_{AccI\,n})^2/(cI_{AccI\,n})^2}{2\sigma^2} \tag{6.16}$$

is searched using the steepest ascent method. Since the unknown parameter β etc. does not appear explicitly, the numerical differentiation is used to obtain the derivatives due to parameters.

6.2. Virtual numerical simulation

6.2.1. Generation of observation model

First, a time series of observation data is generated using the system model and the observation model under the following conditions.

mode 1 When using the number of currently infected people:

$$O_{PrsI\,n} = cI_{PrsI\,n}(1+\varepsilon_n), \quad (n = 1,2,\cdots,N). \tag{6.17}$$

mode 2 When using the daily number of newly infected people:

$$O_{DlyI\,n} = cI_{DlyI\,n}(1+\varepsilon_n), \quad (n = 1,2,\cdots,N). \tag{6.18}$$

mode 3 When using the cumulative number of infected people:

$$O_{Accl\ n} = cI_{Accl\ n}(1+\varepsilon_n), \quad (n=1,2,\cdots,N). \tag{6.19}$$

The data released include the number of new infections, the number of severe cases, and the number of deaths on a daily basis. We consider the following as data. The daily number of newly infected persons is the published data itself, and the cumulative number of infected persons can be calculated from the daily number of newly infected persons, but the current number of infected persons cannot be calculated from the published data. The daily and cumulative numbers of infected people reflect data on those who are infected, but not those who leave the infection. Both of these are reflected in the currently infected person. These things are not a problem in a purely theoretical examination, but they are serious problems in applying the theory to reality.

Table 6.1 List of parameters

Name in Code	Definition	Value
mode	PrsI, DlyI, AccI (Currently, daily and Cumulative number of infected people is used as observation data)	1, 2, 3
Npop	N : Population	1,000
beta1	β : Infection rate (When obs. data generated)	0.4
gam1	γ: Recovery rate (When obs. data generated)	0.04
c1	c : Observation coefft. (When obs. data generated: Ratio to the true data)	0.25
S1[0]	S_0: Initial no of susceptible(When obs. data generated)	997
I1[0]	I_0: Initial no of current infected (When obs. data generated)	3
R1[0]	R_0 : Ini. No of recovered (When obs. data generated)	0
sgmObs1	σ_{Obs} Std. dev. of obs. (When obs. data generated)	0.25
T	Observation period	100
dt	Observation step	0.1
oSkp	Time step of sampling obs. data	10
dlmd	Parameter Differentiation of during numerical diff. (std. val.)	$1.0*10^{-7}$
lmd	Moving step at hill-climbing (std. val.)	$1.0*10^{-7}$
iEnd	No. of convergence cal. at steepest ascent method (std. val.)	30,000
	The initial values of infection rate β, recovery rate γ, initial number of infected people I_0, observation coefficient c, and observation data std. dev. σ_{Obs} when searching for the maximum probability value by the steepest ascent	

	method are 0.8 times the set value at the time of observation data generation. However, I_0 converges slowly, so use the set value.	

6.2.2. Data assimilated numerical simulation

The calculation assumes five unknown parameters: infection rate β, recovery rate γ, initial number of infected people I_0, observation coefficient c, and standard deviation σ_{Obs} of observation data noise. Since the convergence of σ_{Obs} by the steepest ascent method is slow, it would be realistic to make it given. And, it may not be necessary to make the standard deviation σ_{Obs} of the observed data unknown, but it will be supposed to be unknown below.

(A) The setting of virtual infection status

Figure 6.1 shows the virtual infection status according to the settings in Table 6.1. A description of the symbols is in Table 6.1. ρ is the basic reproduction number:

$$\rho = \beta S/(\gamma N). \tag{6.20}$$

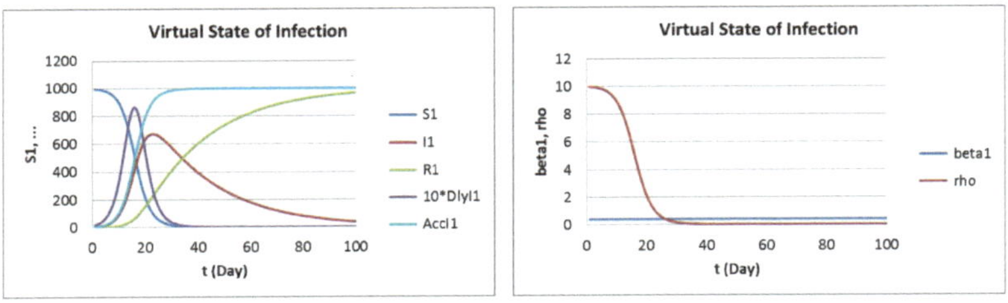

Fig. 6.1. Setting of Assumed Virtual State of Infection.

(B) The effects of the skipping of observation step

In order to ensure the accuracy of integration of the system model, we want to calculate in a 0.1-day time step. Time integration is performed by the Euler method. On the other hand, the virtual observation data is given on a daily basis. Table 6.2 and Fig. 6.2 show the effects when the time step of integration of the system model and the time step of observation are not the same. The parameter I_0 was set to $I_0 = 3$ for the above reasons.

Table 6.2. Effects of Observation Skipping on Data Assimilation.

Parameter	Set Value	Skipping of Time Step for Observation	
		No Skipping	Observation Every 10 Steps
β : Infection Rate	0.4	0.3980	0.3931
γ : Retired Ratio	0.04	0.0366	0.0399
c : Observation Coefft.	0.25	0.2699	0.2769
σ_{Obs}^2 : Variance of Observed Data	0.25^2	0.2716^2	0.2590^2

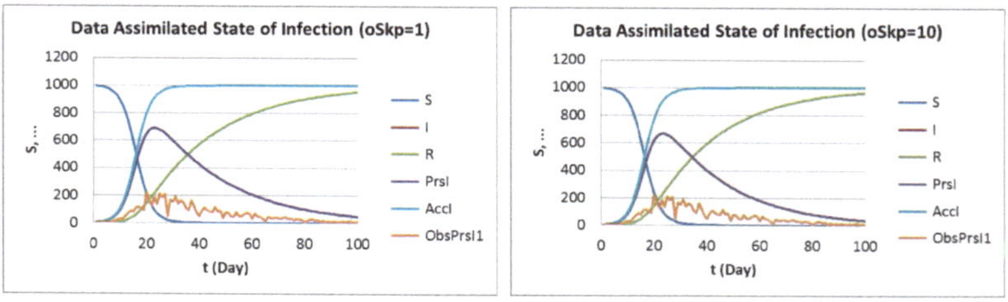

Fig. 6.2. Effects of Observation Skipping on Data Assimilation.

(C) Comparison of Observation Data

The observation data *PrsI* , *DlyI* and *AccI* are compared below. The comparison results are shown in Table 6.3 and Fig. 6.3. In the present calculation, there was a problem with data assimilation by *AccI* . As shown in Table 6.3, the recovery rate γ is not accurate enough. Looking at Fig. 6.3, it is clear that the case of *AccI* is incorrect.

Table 6.3. Effects of Observation mode on Data Assimilation.

Mode	Set Value	*PrsI* : Present Infection	*DlyI* : Daily Infection	*AccI* : Accumulated Infection
β : Infection Rate	0.4	0.3980	0.3982	0.3684
γ: Retired Ratio	0.04	0.0366	0.0402	0.0109
c : Observation Coefft.	0.25	0.2699	0.2653	0.2812
σ_{Obs}^2 : Variance of Observed Data	0.25^2	0.2716^2	0.2584^2	0.2594^2

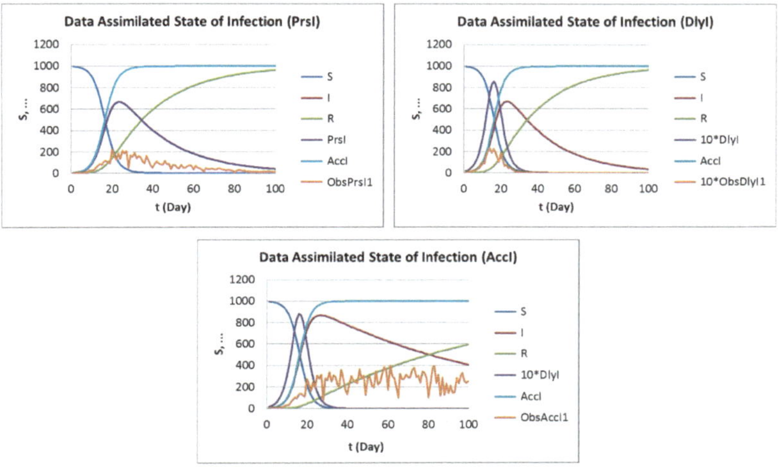

Fig. 6.3. Effects of Observation mode on Data Assimilation.

The daily and cumulative number of infected people reflects data on infected people but does not reflect data on those who leave the infection. *PrsI*, which reflects both the infection rate β and the recovery rate γ, can be said to be the best observation data. However, the observation data cannot be obtained from the currently published data. In the case of *DlyI* and *AccI*, it is better to exclude the recovery rate γ from the estimation and to estimate a fixed value by some method. Table 6.4 and Fig. 6.4 show the results in that case.

Table 6.4. Effects of Fixing Retired Rate γ on Data Assimilation.

Mode	Set Value	*AccI* : Accumulated Infection
β : Infection Rate	0.4	0.3915
γ : Retired Rate	0.04	0.04 (Given)
c : Observation Coefficient	0.25	0.2817
$\sigma_{Obs}{}^2$: Variance of Observed Data	0.25^2	0.2592^2

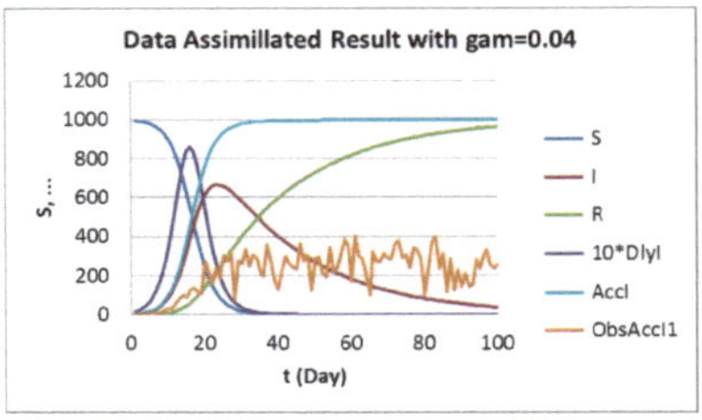

Fig. 6.4. Effects of Fixing Retired Rate γ on Data Assimilation.

(D) Predictive characteristics of data assimilated simulation

Table 6.5 and Fig. 6.5 show the results of data assimilated simulation using the number of currently infected people *PrsI* as the observation data. It can be seen that even if the learning period is short, the infection state for the entire period (100 days) can be predicted fairly accurately.

Table 6.5. Prediction of State of Infection (*PrsI* ; TS: Days for Study).

Mode	Set Value	TS=10	TS=20	TS=50	TS=100
β : Infection Rate	0.4	0.4045	0.4035	0.3998	0.3933
γ : Retired Ratio	0.04	0.0343	0.0342	0.0374	0.0398
c : Observation Coefft.	0.25	0.2463	0.2462	0.2592	0.2763
$\sigma_{Obs}{}^2$: Variance of Observed Data	0.25^2	0.2169^2	0.2171^2	0.2345^2	0.2587^2

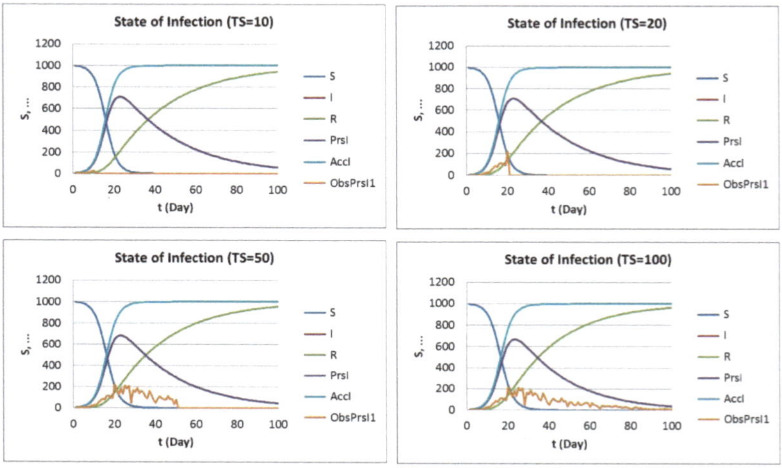

Fig. 6.5. Prediction of State of Infection ($PrsI$; TS: Days for Study)

(E) Effect of social infection control

Infectious diseases can be controlled by social measures such as wearing masks, social distance, and controlling human flow. These effects manifest themselves as effective changes in infection rates. Table 6.6 shows the parameter settings during the numerical simulation, and Fig. 6.6 shows the virtual data observation results and the data assimilated simulation results. The number of currently infected people $PrsI$ is used as observation data.

Fig. 6.6 (a) shows the virtual infection state performed to generate the observation data. It is assumed that the effective infection rate β was reduced from 0.2 to 0.04 with 80% control by two states of emergency declarations made to prevent the spread of infection. Looking at the data assimilated simulation results in Fig. 6.6 (b), it is clear that the virtual infection state is reproduced. However, in this data assimilated simulation, only the infection rate β is set as an unknown variable in order to stably converge the parameters.

In this problem, the infection rate β changes with time, so it is not possible to perform a batch calculation for the entire calculation time T. The calculation is performed assuming that 10 days is one stage and the parameters are constant in each stage.

The programming code is shown in Appendix 6A.

Table 6.6. Parameters for Numerical Simulation

Name in Code	Definition	Value
mode	$PrsI$ (Currently number of infected people is used as observation data)	1
Npop	N : Population	32,000
beta1	β : Infection rate (When obs. data generated)	0.2
gam1	γ: Recovery rate (When obs. data generated)	0.05
c1	c : Observation coefft. (When obs. data generated: Ratio to the true data)	0.7
S1[0]	S_0: Initial no of susceptible(When obs. data generated)	31,998
I1[0]	I_0: Initial no of current infected (When obs. data generated)	2
R1[0]	R_0 : Ini. No of recovered (When obs. data generated)	0
sgmObs1	σ_{Obs} : Std. dev. of obs. (When obs. data generated)	0.25
T	Observation period	360
dt	Observation step	0.1
oSkp	Time step of sampling obs. data	10
dlmd	Parameter Differentiation of during numerical diff.	$1.0*10^{-8}$

	(std. val.)	
lmd	Moving step at steepest ascent (std. val.)	$2.0*10^{-7}$
iEnd	No. of convergence cal. at steepest ascent (std. val.)	50,000
	The entire period (360 days) is calculated by dividing it into 36 stages where 1 stage means 10 days. The maximum probability is searched by the steepest ascent method, using only the infection rate β as an unknown parameter to make the convergence stable. The initial value of the infection rate is 0.8 times the set value at the generation of the observation data.	

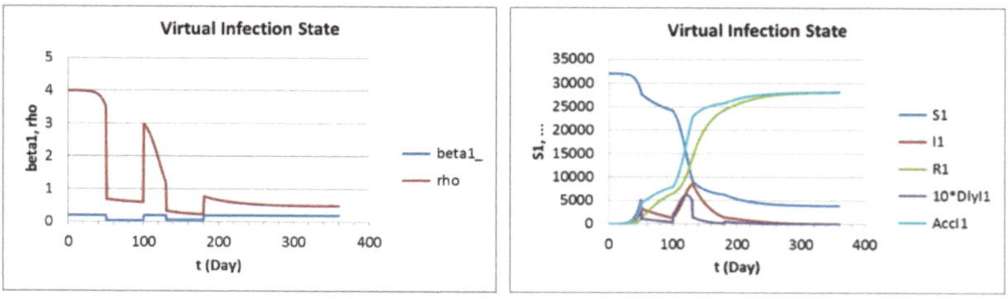

(a) Virtual Infection State

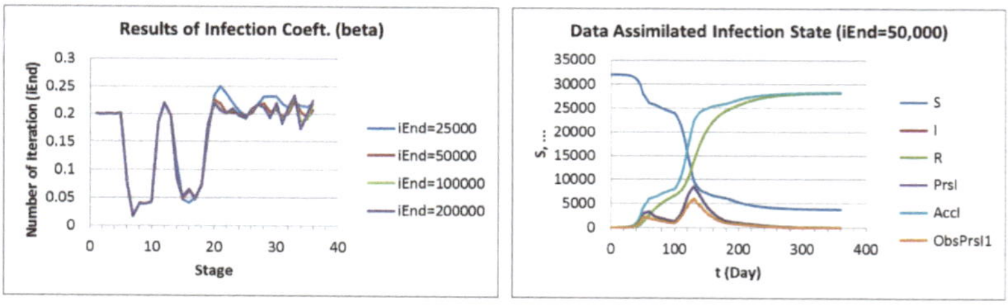

(b) Data Assimilatd Infection State

Fig. 6.6. Data Assimilatd Infection State.

6.3. Data assimilated numerical simulation using real data

Unlike the case of virtual data assimilation using virtual observation data, real data assimilated simulation using actually observed real data is not straightforward. The study has just begun, although it seems to contain a variety of issues.

In the handling of COVID-19 in Japan, the persons found infected are quarantined and the infectivity is contained, so that amount must be subtracted from the presently infected persons I in the calculation of the newly infected person. Applying SIR theory to real data is problematic. However, if the spread of infection is not sufficiently isolated, the outline can be grasped by using SIR theory for the time being. As a result, the infection rate will be underestimated.

Although the observation data of Tokyo was used, the SIR equation, which is the mean-field theory, cannot be used as it is. The population of Tokyo is about 14 million, but the effective infection opportunity population involved in the infection at the time is only a small part of it. It increases with the spread of infection. In the following calculation, it is considered that the infection spreads in a ring shape at a constant rate, and it is considered that the infection increases with a linear function of time. As a result of trial and error, 16,000 people at 120 days and 200,000 people at 370 days. The numerical value connecting the two points with a straight line was taken as N.

As an observation model for real data, the following expression was used:

$$O_{AccI\,n} = cI_{AccI\,n} + \varepsilon_n, \quad (n = 1, 2, \cdots, N). \tag{6.21}$$

The parameters used in the data assimilated simulation are shown in Table 6.7, the data observation results are shown in Fig. 6.7, and the assimilated simulation results are shown in Fig. 6.8.

Until now, the phenomenon of infection wave could not be rationally reproduced in the calculation. If people's social activities are changed by a state of emergency, etc., the infection rate will change over time, causing wave motion. The infection waves have been successfully reproduced.

Comparing the 120-day analysis results in Fig. 6.8 with the 370-day analysis results, there is a difference in the results for the first 120 days, which should be the same. This is because, when applying the mean-field theory, the number of people who have the chance of infection is different from 16,000 and 200,000. The solution to this problem is a future problem, but it can be said that the current calculation results do not deny the validity of this calculation method.

In this calculation, only the infection rate β is an unknown parameter. In order to make other parameters unknown, at least observation data that reflects the effect of the recovery rate γ will be required. Since only the infection rate β is unknown, it seems that the temporal changes in other parameters are not sufficiently reflected in the changes in the infection rate β.

In addition, the infection rate that appears in the actual infection phenomenon is not a pathologically defined infection rate with strictly defined conditions, but an effective infection rate that takes into account social impacts such as wearing masks, social distance, and human flow. The infection rate in Fig. 6.8 is such, and the effect of artificially suppressing infection by declaring an emergency is reflected some extent.

Table 6.7 List of Parameters.

Name in Code	Definition	Value
mode	*AccI* (Cumulative number of infected people is used as observation data)	3
Npop	N: Effective Infective opportunity Population (Given	16,000–

	at each stage)	200,000
beta1	β : Infection rate (Initial value for convergence calculation)	0.2
gam1	γ: Recovery rate (Given at the beginning)	0.05
c1	c : Observation coefft. (Given at the beginning)	0.6
S1[0]	S_0 : Initial no of susceptible(Given at the beginning)	31,998
I1[0]	I_0 : Initial no of current infected (Given at the beginning)	15,999 – 199,999
R1[0]	R_0 : Ini. Number of recovered (Given at the beginning)	0
sgmObs1	σ_{Obs} : Std. dev. of obs. (Given at each stage)	240–4,000
T	Observation period	370
dt	Observation step	1
oSkp	Time step of sampling obs. data	1
dlmd	Denominator of numerical differentiation (std. val.)	$1.0*10^{-8}$
lmd	Moving step at steepest ascent (std. val.)	$2.0*10^{-7}$
iEnd	No. of convergence cal. at steepest ascent (std. val.)	32,000
	The entire period (370 days) is calculated by dividing it into 37 stages where 1 stage means 10 days. The maximum probability is searched by the steepest ascent method, using only the infection rate β as an unknown parameter to make the convergence stable. The initial value of the infection rate is 0.8 times the set value at the generation of the observation data.	

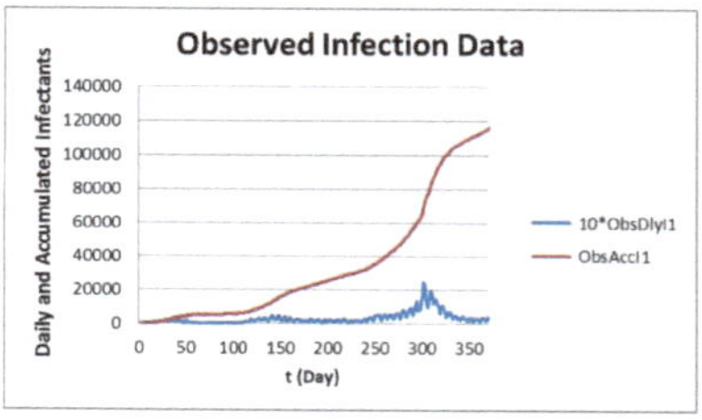

Fig. 6.7. Observed Infection Data.

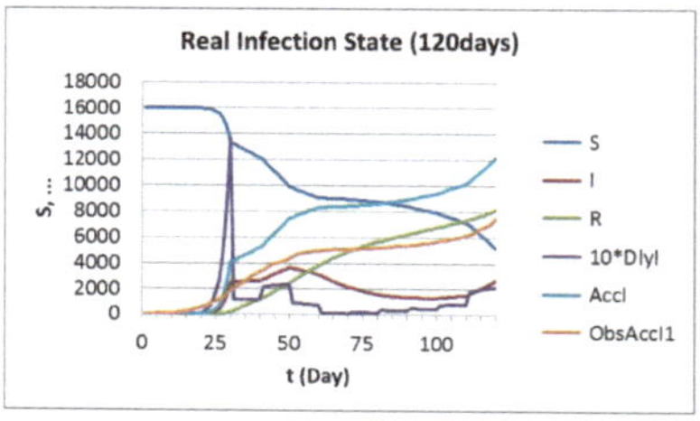

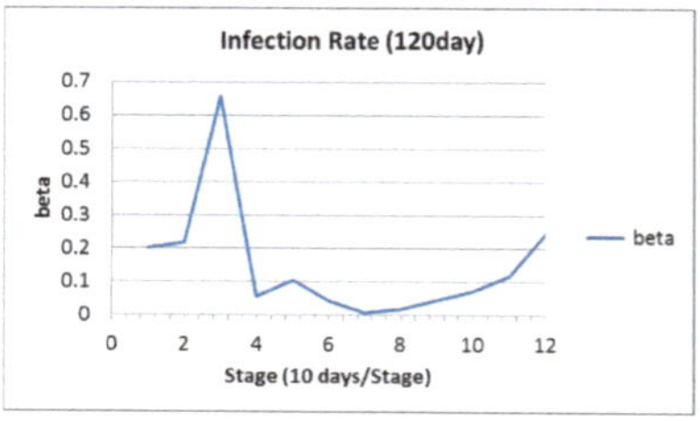

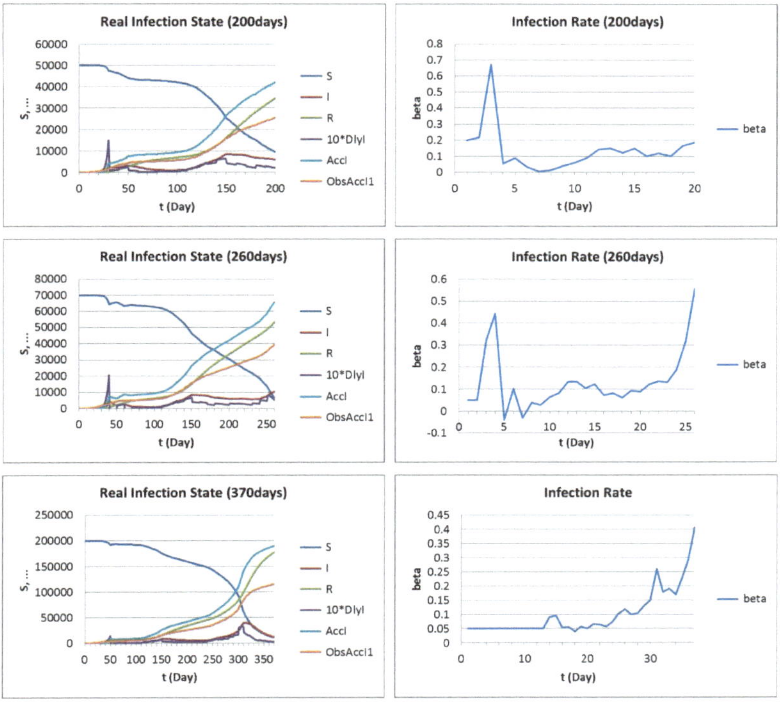

Fig. 6.8. Data Assimilation Results Using Observation Data of Tokyo (2020.03.13 - 2021.03.17)

6.4. Discussions

The new coronavirus infection (COVID-19) is rampant. Even now, more than a year after the start of the infection, the infection has not yet shown signs of ending. It can be said to be an infection that remains in history.

The most troublesome part of this infection is that it not only takes several days from infection to onset, but it also infects for several days even in the not-onset state. Therefore, a considerable number of infected persons with infectivity are left unchecked. Therefore, even if the infection status is simulated by the SIR equation [1-3], the true values of the infection parameters and the number of infected persons cannot be grasped.

However, it may be possible to estimate the true value by combining it with the observation equation. In short, a state-space model or a data assimilation framework is considered to be effective. The parameters of the SIR equation, such as the daily number of new infections, infection rate, and recovery rate, are unknown but embedded in the observed values. Unknown parameters must match the observed values. Wouldn't it be possible to determine these unknowns by combining them using Bayesian inference or the method of least squares? This effectiveness was confirmed by the numerical results somewhat.

Numerical calculations were performed not only when the observation data was artificially generated, but also when the actually published observation data was used. In the former case, consistent results were obtained for all observed data of the current number of infected persons, the number of newly infected persons on a daily basis, and the cumulative number of infected persons. Regarding the latter case, the daily number of newly infected persons and the cumulative number of infected persons were used as observation data, and in this case as well, consistent results were obtained regardless of which observation data was used.

However, the daily number of newly infected persons and the cumulative number of infected persons reflect the infection rate β, but do not reflect the recovery rate. As the observation data required for data assimilation, good data that reflects the recovery rate γ is absolutely necessary.

For virus mutant strains, the current concept of time-varying parameters is sufficient, but for the effect of the vaccines, it is necessary to subtract the vaccinated persons from the infected persons S. In future studies, the author would also like to use the infection equation including the effect of the vaccine.

However, efforts to assimilate data that combine the mathematical theory and observational data of the time evolution of infectious diseases have only just begun, and many issues have not been fully considered. The current method of collecting and organizing statistical data on infectious diseases is not premised on being used for data assimilation. On the premise of introducing data assimilation, it will be necessary to rethink the way statistical data should be.

Acknowledgement

The author would like to thank Professor Emeritus Hiroyuki Kajiwara of Kyushu University for providing the actual data of the new coronavirus infection (COVID-19).

REFERENCES 6

[1] W. 0. Kermack, A. G. McKendrick, A Contribution to the Mathematical Theory of Epidemics, 1927, Proceedings of the Royal Society A (1927). https://doi.org/10.1098/rspa.1927.0118.

[2] H. Isshiki, M. Namiki, T. Kinoshita, R. Yano : Effective Infection Opportunity Population (EOIP) Hypothesis in Applying SIR Infection Theory, cornell arXive, arXiv:2009.01837 (2020).

https://arxiv.org/search/?query=Hiroshi+Isshiki&searchtype=all&source=headr

[3] G. Kitagawa, Use of a State Space Model in Time Series Analysis, Proceedings of the Institute of Statistical Mathematics, Vol. 67, No. 2 (2019) 181–192, in Japanese.
https://www.ism.ac.jp/editsec/toukei/pdf/67-2-181.pdf

[4] K. Fukaya, Time series analysis by state space models and its application in ecology, Japanese Journal of Ecology, Vol. 66, No. 2 (2016), 375-389 in Japanese.
https://doi.org/10.18960/seitai.66.2_375

[5] K. Law, A. Stuart, K. Zygalakis, Data Assimilation: A Mathematical Introduction, Springer (2015).

Appendix 6A Code for data assimilatd infection state

"Microsoft C/C++ Compiler Version 17.00.50727.1 for x86" and "Microsoft Linker Version 11.00.50727" were used for compile and link (in command window; cl source_file_name.c).

(1) Programing code: EstTimeSeq0SSMZ4TmDPdMtClmbChgPrmYFineTime.c

```
/* ------------------------------------------------------------------------ */
/*                                                                          */
/* File Name: EstTimeSeq0.c                                 2020.04.30-2020.04.30 */
/* File Name: EstTimeSeq0SSM.c                              2021.03.20-2021.03.21 */
/* File Name: EstTimeSeq0SSMX.c                             2021.03.20-2021.03.23 */
/* File Name: EstTimeSeq0SSMY.c                             2021.03.24-2021.03.24 */
/* File Name: EstTimeSeq0SSMZ.c                             2021.03.24-2021.03.24 */
/* File Name: EstTimeSeq0SSMZ1.c                            2021.03.24-2021.03.26 */
/* File Name: EstTimeSeq0SSMZ2.c                            2021.03.26-2021.03.27 */
/* File Name: EstTimeSeq0SSMZ3.c                            2021.03.27-2021.03.28 */
/* File Name: EstTimeSeq0SSMZ4TmDPd.c                       2021.03.28-2021.04.01 */
/* File Name: EstTimeSeq0SSMZ4TmDPdMtClmb.c                 2021.04.08-2021.04.09 */
/* File Name: EstTimeSeq0SSMZ4TmDPdMtClmbChgPrm.c           2021.04.10-2021.04.10 */
/* File Name: EstTimeSeq0SSMZ4TmDPdMtClmbChgPrm.c           2021.04.10-2021.04.13 */
/* File Name: EstTimeSeq0SSMZ4TmDPdMtClmbChgPrmX.c          2021.04.10-2021.04.14 */
/* File Name: EstTimeSeq0SSMZ4TmDPdMtClmbChgPrmY.c          2021.04.14-2021.04.18 */
/* File Name: EstTimeSeq0SSMZ4TmDPdMtClmbChgPrmYFineTime.c  2021.04.19-2021.05.01 */
/* File Name: EstTimeSeq0SSMZ4TmDPdMtClmbChgPrmYFineTime.c  2022.05.24-2022.05.24 */
/*                                                                          */
/*    Estimation Time Sequence ... Data Assisted Infection State            */
/*                                                                          */
/* ------------------------------------------------------------------------ */

#include <stdio.h>
#include <stdlib.h>
#include <string.h>
#include <math.h>

#define PI      3.14159265

void main();
void pushKey();
```

```
void func(double x,double y[],double f[]);

double Uniform( void );                    // uniform random number [0,1]
double rand_normal(double,double);         // normal random number

double normal(double,double,double);       // normal distribution

double Pnml_I(int n, double c, double myu, double sgm);    //
double logPnml_I(int n, double c, double myu, double sgm); //

void SIRsol(int, int, double, double);                     // solution of SIR equation

/* ----------------------------------------------------------------- */

//int P;                  // number of candidates

long NS;                  // number of time steps for study
long NE;                  // number of time steps for interpolation + prediction

int NStg;                 // number of stage
int WStg;                 // width of stage

int NBgn;                 // number of time steps for begin of study
int NEnd;                 // number of time steps for end of study

double T;                 // total time in days
double t;                 // time in days
double dt;                // time interval

int Case;                 // Case 0: flat, 1: down, 2: up, 3: down up

int mode;                 // mode of observation

int Npop;                 // population

double beta1;             // given coefficient of infection
double beta1_[10001];     // given coefficient of infection
double gam1;              // given coefficient of retrieved
double c1;                // given coefficient of retrieved
```

```
double S1[10001];                   // number of susceptible people
double I1[10001];                   // number of infective people
double R1[10001];                   // number of retrieved people (recovered+daed)
double PrsI1[10001];                // number of present infective people
double DlyI1[10001];                // number of daily infective people
double AccI1[10001];                // number of accumulated infective people

double ObsPrsI1[10001];             // observation: number of present infective people
double sgmObsPrsI1;                 // parameter in state space representation for obs. data
generation
double sgmObsPrsI2;                 // parameter in state space representation for prediction
double sgmObsPrsI[41];              // parameter in state space representation
double epsObsPrsI1[10001];          // observation noise ... random number following normal
distributin

double ObsDlyI1[10001];             // observation: number of daily infective people
double sgmObsDlyI1;                 // parameter in state space representation for obs. data
generation
double sgmObsDlyI2;                 // parameter in state space representation for prediction
double sgmObsDlyI[41];              // parameter in state space representation
double epsObsDlyI1[10001];          // observation noise ... random number following normal
distributin

double ObsAccI1[10001];             // observation: number of acumulative infective people
double sgmObsAccI1;                 // parameter in state space representation for obs. data
generation
double sgmObsAccI2;                 // parameter in state space representation for prediction
double sgmObsAccI[41];              // parameter in state space representation
double epsObsAccI1[10001];          // observation noise ... random number following normal
distributin

double lmd;                         // step of mountain climbing
double dlmd;                        // difference in derivative
int iEnd;                           // end of mount climbing
int iSkp;                           // skip print lines for convergence
int nSkp;                           // skip print lines for process results
int oSkp;                           // skip observation

 double beta_[41];                  // parameter beta
```

```
double gam_[41];                // parameter gam
double c_[41];                  // parameter c
double sgm_[41];                // parameter sgm

double beta_Rcd[41];            // parameter beta_ ... record
double gam_Rcd[41];             // parameter gam_ ... record
double c_Rcd[41];               // parameter c_ ... record
double sgm_Rcd[41];             // parameter sgm_ ... record
double RProb_Rcd[41];           // parameter logRProb ... record
double logRProb_Rcd[41];        // parameter logRProb ... record

double beta[41];                // coefficient of infection
double gam[41];                 // coefficient of retrieved
double c[41];                   // parameter in state space representation
double S[10001];                // number of susceptible people
double I[10001];                // number of infective people
double R[10001];                // number of retrieved people (recovered+daed)
double PrsI[10001];             // number of present infective people
double DlyI[10001];             // number of daily infective people
double AccI[10001];             // number of accumulated infective people

double R0[10001];               // epidemological threshold ... function of time

FILE *fp_inp;                   // pointer of input file
FILE *fp_out;                   // pointer of output file

char InputDataFile[80];         // input file name
char OutputDataFile[80];        // output file name

char buf[5000];
double vec1[1001];              // temporary vector

double logRProb_;               // log of RProb_
double logRProb_beta;           // derivative of logRProb_ wrt beta
double logRProb_gam;            // derivative of logRProb_ wrt gam
double logRProb_c;              // derivative of logRProb_ wrt c
double logRProb_sgm;            // derivative of logRProb_ wrt sgm

int PrtCtrl;                    // print control ... 1: print
```

```
int betacon;                        // control infectant
int gamcon;                         // control infectant
//int Icon;                           // control infectant
int ccon;                           // control infectant
int sgmcon;                         // control infectant

double betaini;                     // initial beta
double gamini;                      // initial gam
//double Iini;                        // initial I
double cini;                        // initial c
double sgmini;                      // initial sgm

/* ------------------------------------------------------------ */

void main()
{
    int p1, p2;
    int i, j, k, n;
    int p, q, r, s;
    int stage;
    double sum, tmp;

    // Input file
    sprintf(InputDataFile, "EstTimeSeqOSSMZ4TmDpdMtClmbChgPrmYFineTime_inp.dat");

    if ((fp_inp = fopen(InputDataFile, "r")) == NULL) {
        printf("Failed in Reading Input Data File! ... %s¥n", InputDataFile);
        exit(1);
    }

    // Output file
    sprintf(OutputDataFile, "EstTimeSeqOSSMZ4TmDpdMtClmbChgPrmYFineTime_out.csv");

    if ((fp_out = fopen(OutputDataFile, "w")) == NULL) {
        printf("Failed in Reading Output Data File! ... %s¥n", OutputDataFile);
        exit(1);
    }
```

```
fscanf(fp_inp, "%s %d", buf, &Case);
fscanf(fp_inp, "%s %d", buf, &mode);
fscanf(fp_inp, "%s %d", buf, &Npop);
fscanf(fp_inp, "%s %lf", buf, &beta1);
fscanf(fp_inp, "%s %lf", buf, &gam1);
fscanf(fp_inp, "%s %lf", buf, &c1);
fscanf(fp_inp, "%s %lf", buf, &S1[0]);
fscanf(fp_inp, "%s %lf", buf, &I1[0]);
fscanf(fp_inp, "%s %lf", buf, &R1[0]);
if (mode == 1) {
    fscanf(fp_inp, "%s %lf", buf, &sgmObsPrsI1);
    fscanf(fp_inp, "%s %lf", buf, &sgmObsPrsI2);
}
else if (mode == 2) {
    fscanf(fp_inp, "%s %lf", buf, &sgmObsDlyI1);
    fscanf(fp_inp, "%s %lf", buf, &sgmObsDlyI2);
}
else if (mode == 3) {
    fscanf(fp_inp, "%s %lf", buf, &sgmObsAccI1);
    fscanf(fp_inp, "%s %lf", buf, &sgmObsAccI2);
}
else {
    fprintf(fp_out, "mode should be 1, 2 or 3¥n");
    exit(0);
}
fscanf(fp_inp, "%s %d", buf, &betacon);
fscanf(fp_inp, "%s %d", buf, &gamcon);
fscanf(fp_inp, "%s %d", buf, &ccon);
fscanf(fp_inp, "%s %d", buf, &sgmcon);
fscanf(fp_inp, "%s %lf", buf, &betaini);
fscanf(fp_inp, "%s %lf", buf, &gamini);
fscanf(fp_inp, "%s %lf", buf, &cini);
fscanf(fp_inp, "%s %lf", buf, &sgmini);
fscanf(fp_inp, "%s %d", buf, &NS);
fscanf(fp_inp, "%s %d", buf, &NE);
fscanf(fp_inp, "%s %d", buf, &NStg);
fscanf(fp_inp, "%s %d", buf, &WStg);
fscanf(fp_inp, "%s %lf", buf, &dt);
```

```
fscanf(fp_inp, "%s %lf", buf, &dlmd);
fscanf(fp_inp, "%s %lf", buf, &lmd);
fscanf(fp_inp, "%s %d", buf, &iEnd);
fscanf(fp_inp, "%s %d", buf, &iSkp);
fscanf(fp_inp, "%s %d", buf, &nSkp);
fscanf(fp_inp, "%s %d", buf, &oSkp);
fscanf(fp_inp, "%s %d", buf, &PrtCtrl);


printf("Case        = %d¥n", Case);
printf("mode        = %d¥n", mode);
printf("Npop        = %d¥n", Npop);
printf("beta1       = %12.6f¥n", beta1);
printf("gam1        = %12.6f¥n", gam1);
printf("c1          = %12.6f¥n", c1);
printf("S1{0]       = %12.6f¥n", S1[0]);
printf("I1{0]       = %12.6f¥n", I1[0]);
printf("R1{0]       = %12.6f¥n", R1[0]);
if (mode == 1) {
    printf("sgmObsPrsI1 = %12.6f¥n", sgmObsPrsI1);
    printf("sgmObsPrsI2 = %12.6f¥n", sgmObsPrsI2);
}
else if (mode == 2) {
    printf("sgmObsDlyI1 = %12.6f¥n", sgmObsDlyI1);
    printf("sgmObsDlyI2 = %12.6f¥n", sgmObsDlyI2);
}
else if (mode == 3) {
    printf("sgmObsAccI1 = %12.6f¥n", sgmObsAccI1);
    printf("sgmObsAccI2 = %12.6f¥n", sgmObsAccI2);
}
else {
    fprintf(fp_out, "mode should be 1, 2 or 3¥n");
    exit(0);
}
printf("betacon     = %d¥n", betacon);
printf("gamcon      = %d¥n", gamcon);
printf("ccon        = %d¥n", ccon);
printf("sgmcon      = %d¥n", sgmcon);
printf("betaini     = %12.6f¥n", betaini);
```

```
printf("gamini      = %12.6f¥n", gamini);
printf("cini        = %12.6f¥n", cini);
printf("sgmini      = %12.6f¥n", sgmini);
printf("NS          = %d¥n", NS);
printf("NE          = %d¥n", NE);
printf("NStg        = %d¥n", NStg);
printf("WStg        = %d¥n", WStg);
printf("dt          = %12.6f¥n", dt);
printf("dlmd        = %12.9f¥n", dlmd);
printf("lmd         = %12.9f¥n", lmd);
printf("iEnd        = %d¥n", iEnd);
printf("iSkp        = %d¥n", iSkp);
printf("nSkp        = %d¥n", nSkp);
printf("oSkp        = %d¥n", oSkp);
printf("PrtCtrl     = %d¥n", PrtCtrl);

fprintf(fp_out, "Case =, %d¥n", Case);
fprintf(fp_out, "Npop =, %d¥n", mode);
fprintf(fp_out, "Npop =, %d¥n", Npop);
fprintf(fp_out, "beta1 =, %12.6f¥n", beta1);
fprintf(fp_out, "gam1 =, %12.6f¥n", gam1);
fprintf(fp_out, "c1 =, %12.6f¥n", c1);
fprintf(fp_out, "S1{0] =, %12.6f¥n", S1[0]);
fprintf(fp_out, "I1{0] =, %12.6f¥n", I1[0]);
fprintf(fp_out, "R1{0] =, %12.6f¥n", R1[0]);
if (mode == 1) {
    fprintf(fp_out, "sgmObsPrsI1 =, %12.6f¥n", sgmObsPrsI1);
    fprintf(fp_out, "sgmObsPrsI2 =, %12.6f¥n", sgmObsPrsI2);
}
else if (mode == 2) {
    fprintf(fp_out, "sgmObsDlyI1 =, %12.6f¥n", sgmObsDlyI1);
    fprintf(fp_out, "sgmObsDlyI2 =, %12.6f¥n", sgmObsDlyI2);
}
else if (mode == 3) {
    fprintf(fp_out, "sgmObsAccI1 =, %12.6f¥n", sgmObsAccI1);
    fprintf(fp_out, "sgmObsAccI2 =, %12.6f¥n", sgmObsAccI2);
}
else {
    fprintf(fp_out, "mode should be 1, 2 or 3¥n");
```

```
    exit(0);
}
fprintf(fp_out, "betacon =, %d¥n", betacon);
fprintf(fp_out, "gamcon =,  %d¥n", gamcon);
fprintf(fp_out, "ccon =, %d¥n", ccon);
fprintf(fp_out, "sgmcon =, %d¥n", sgmcon);
fprintf(fp_out, "betaini =, %12.6f¥n", betaini);
fprintf(fp_out, "gamini =,  %12.6f¥n", gamini);
fprintf(fp_out, "cini =, %12.6f¥n", cini);
fprintf(fp_out, "sgmini =, %12.6f¥n", sgmini);
fprintf(fp_out, "NS =, %d¥n", NS);
fprintf(fp_out, "NE =, %d¥n", NE);
fprintf(fp_out, "NStg =, %d¥n", NStg);
fprintf(fp_out, "WStg =, %d¥n", WStg);
fprintf(fp_out, "dt =, %12.6f¥n", dt);
fprintf(fp_out, "dlmd =, %12.9f¥n", dlmd);
fprintf(fp_out, "lmd =, %12.9f¥n", lmd);
fprintf(fp_out, "iEnd =, %d¥n", iEnd);
fprintf(fp_out, "iSkp =, %d¥n", iSkp);
fprintf(fp_out, "nSkp =, %d¥n", nSkp);
fprintf(fp_out, "oSkp =, %d¥n", oSkp);
fprintf(fp_out, "PrtCtrl =, %d¥n", PrtCtrl);
fprintf(fp_out, "¥n");

pushKey();

T = (NS+0.0)*dt;
t = 0.0;


if (Case == 0)     // flat
    for (n = 1; n <= NE; n++) {
        if (n <= 50/dt)
            beta1_[n] = beta1*1.0;
        else
            beta1_[n] = beta1*1.0;
    }
else if (Case == 1)     // down step
    for (n = 1; n <= NE; n++) {
```

```
            if (n <= 50/dt)
               beta1_[n] = beta1*1.0;
            else
               beta1_[n] = beta1*0.2;
        }
    else if (Case == 2/dt)     // up step
        for (n = 1; n <= NE; n++) {
            if (n <= 50/dt)
               beta1_[n] = beta1*1.0;
            else
               beta1_[n] = beta1*1.5;
        }
    else if (Case == 3)     // down up
        for (n = 1; n <= NE; n++) {
            if (n <= 50/dt)
               beta1_[n] = beta1*1.0;
            else if (n >= 51/dt && n <= 100/dt)
               beta1_[n] = beta1*0.2;
            else
               beta1_[n] = beta1*1.0;
        }
    else if (Case == 4)     // down up
        for (n = 1; n <= NE; n++) {
            if (n <= 50/dt)
               beta1_[n] = beta1*1.0;
            else if (n >= 51/dt && n <= 100/dt)
               beta1_[n] = beta1*0.2;
            else if (n >= 101/dt && n <= 130/dt)
               beta1_[n] = beta1*1.0;
            else if (n >= 131/dt && n <= 180/dt)
               beta1_[n] = beta1*0.3;
            else
               beta1_[n] = beta1*1.0;
        }
    else {
        fprintf(fp_out, "Case shoude be 0-4¥n");
        exit(0);
    }
```

```
    fprintf(fp_out, "n, t, S1, I1, R1, beta1_, rho¥n");
    for (n = 1; n <= NS; n++) {
        t = (n+0.0)*dt;

        S1[n] = S1[n-1] -beta1_[n]/(Npop+0.0)*S1[n-1]*I1[n-1]*dt;
        I1[n] = I1[n-1] + (beta1_[n]/(Npop+0.0)*S1[n-1]*I1[n-1] - gam1*I1[n-1])*dt;
        R1[n] = R1[n-1] + gam1*I1[n-1]*dt;

        if (n % nSkp == 0)
            fprintf(fp_out, "%d, %12.6f, %12.6f, %12.6f, %12.6f, %12.6f, %12.6f¥n",
                    n, t, S1[n], I1[n], R1[n], beta1_[n], beta1_[n]*S1[n]/Npop/gam1);
    }
    fprintf(fp_out, "¥n");


    // epsObsPrsI1[n] ... random number following normal distribution ... present infectants
    // epsObsDlyI1[n] ... random number following normal distribution ... daily infectants
    // epsObsAccI1[n] ... random number following normal distribution ... accumulated
infectnts

    fprintf(fp_out, "n, t, epsObsI¥n");


    for (n = 1; n <= NS; n++) {
        t = (n+0.0)*dt;

        if (mode == 1) {
            epsObsPrsI1[n] = rand_normal(0.0,sgmObsPrsI1);
            if (PrtCtrl == 1)
                fprintf(fp_out, "%d, %12.6f, %12.6f¥n", n, t, epsObsPrsI1[n]);
        }
        else if (mode == 2) {
            epsObsDlyI1[n] = rand_normal(0.0,sgmObsDlyI1);
            if (PrtCtrl == 1)
                fprintf(fp_out, "%d, %12.6f, %12.6f¥n", n, t, epsObsDlyI1[n]);
        }
        else if (mode == 3) {
            epsObsAccI1[n] = rand_normal(0.0,sgmObsAccI1);
```

```
            if (PrtCtrl == 1)
                fprintf(fp_out, "%d, %12.6f, %12.6f¥n", n, t, epsObsAccI1[n]);
        }
        else
            fprintf(fp_out, "mode should be 1, 2 or 3¥n");

    }
    fprintf(fp_out, "¥n");

    fprintf(fp_out, "n, t, S1, I1, R1, 10*DlyI1, AccI1, beta1_, rho¥n");
    AccI1[0] = I1[0];
    for (n = 1; n <= NS; n++) {
        t = (n+0.0)*dt;

        PrsI1[n] = I1[n];
        DlyI1[n] = beta1_[n]/(Npop+0.0)*I1[n]*S1[n];
        AccI1[n] = AccI1[n-1] + DlyI1[n]*dt;

        if (mode == 1) {
            ObsPrsI1[n] = c1*PrsI1[n]*(1.0+epsObsPrsI1[n]);
        }
        else if (mode == 2) {
            ObsDlyI1[n] = c1*DlyI1[n]*(1.0+epsObsDlyI1[n]);
        }
        else if (mode == 3) {
            ObsAccI1[n] = c1*AccI1[n]*(1.0+epsObsAccI1[n]);
        }
        else
            fprintf(fp_out, "mode should be 1, 2 or 3¥n");

        if (n % nSkp == 0)
            fprintf(fp_out,
"%d, %12.6f, %12.6f, %12.6f, %12.6f, %12.6f, %12.6f, %12.6f, %12.6f¥n",
                    n, t, S1[n], I1[n], R1[n], 10.0*DlyI1[n], AccI1[n], beta1_[n],
beta1_[n]*S1[n]/Npop/gam1);
    }
    fprintf(fp_out, "¥n");
```

```
    fprintf(fp_out, "¥n");
    /// Mountain Climbing ... numerical differentiation
    fprintf(fp_out, "/// MOUNTAIN CLIMBING¥n¥n");

    // Study...Mountain Climbing, n: from 1 to NE ... numerical differentiation
    fprintf(fp_out, "// Study...Mountain Climbing, n: from 1 to NE ... numerical
differentiation¥n");
    fprintf(fp_out, "¥n");
    fprintf(fp_out, "¥n");



    stage = 1;
    beta_[stage] = beta1;
    gam_[stage] = gam1;
    c_[stage] = c1;
    if (mode == 1)
        sgm_[stage] = sgmObsPrsI2;
    else if (mode == 2)
        sgm_[stage] = sgmObsDlyI2;
    else if (mode == 3)
        sgm_[stage] = sgmObsAccI2;
    else {
        fprintf(fp_out, "mode should be 1, 2 or 3¥n");
        exit(0);
    }


    fprintf(fp_out, "¥n¥n¥n**********, beta-1, **********¥n");

    for (stage = 1; stage <= NStg; stage++) {
        NBgn = (stage-1)*WStg+1;
        NEnd = stage*WStg;


        if (stage >= 2) {
            beta_[stage] = beta_[stage-1];
            gam_[stage] = gam_[stage-1];
            c_[stage] = c_[stage-1];
```

```
        sgm_[stage] = sgm_[stage-1];
    }


    fprintf(fp_out, "*****, stage =, %d, NBgn =, %d, NEnd = %d *****¥n", stage, NBgn,
NEnd);

    fprintf(fp_out, "i, beta_, gam_, c_, sgm_, logRProb, logRProb_beta, logRProb_gam,
logRProb_c, logRProb_sgm¥n");

    for (i = 1; i <= iEnd; i++) {

        S[0] = S1[0];
        I[0] = I1[0];
        R[0] = R1[0];
        PrsI[0] = I[0];
        DlyI[0] = 0.0;
        AccI[0] = I[0];

        SIRsol(NBgn, NEnd, beta_[stage], gam_[stage]);
        logRProb_ = 0.0;
        for (n = NBgn; n <= NEnd; n++) {
            if (n % oSkp == 0)
                logRProb_ += logPnml_I(n, c_[stage], 0.0, sgm_[stage]);
        }


        if (betacon == 1) {
            SIRsol(NBgn, NEnd, beta_[stage]+dlmd, gam_[stage]);
            logRProb_beta = 0.0;
            for (n = NBgn; n <= NEnd; n++)
                if (n % oSkp == 0)
                    logRProb_beta += logPnml_I(n, c_[stage], 0.0, sgm_[stage]);
            logRProb_beta = (logRProb_beta-logRProb_)/dlmd;
        }

        if (gamcon == 1) {
            SIRsol(NBgn, NEnd, beta_[stage], gam_[stage]+dlmd);
            logRProb_gam = 0.0;
```

```
        for (n = NBgn; n <= NEnd; n++)
            if (n % oSkp == 0)
                logRProb_gam += logPnml_I(n, c_[stage], 0.0, sgm_[stage]);
        logRProb_gam = (logRProb_gam-logRProb_)/dlmd;
    }

    if (ccon == 1) {
        SIRsol(NBgn, NEnd, beta_[stage], gam_[stage]);
        logRProb_c = 0.0;
        for (n = NBgn; n <= NEnd; n++)
        if (n % oSkp == 0)
            logRProb_c += logPnml_I(n, c_[stage]+dlmd, 0.0, sgm_[stage]);
        logRProb_c = (logRProb_c-logRProb_)/dlmd;
    }


    if (sgmcon == 1) {
        SIRsol(NBgn, NEnd, beta_[stage], gam_[stage]);
        logRProb_sgm = 0.0;
        for (n = NBgn; n <= NEnd; n++)
            if (n % oSkp == 0)
                logRProb_sgm += logPnml_I(n, c_[stage], 0.0, sgm_[stage]+dlmd);
        logRProb_sgm = (logRProb_sgm-logRProb_)/dlmd;
    }

    if (betacon == 1)
        beta_[stage] += logRProb_beta*lmd;
    if (gamcon == 1)
        gam_[stage] += logRProb_gam*lmd;
    if (ccon == 1)
        c_[stage] += logRProb_c*lmd;
    if (sgmcon == 1)
        sgm_[stage] += logRProb_sgm*lmd;

    if (beta_[stage] < 0.0)
        beta_[stage] = 0.0;
    if (gam_[stage] < 0.0)
        gam_[stage] = 0.0;
```

```
            if (i % iSkp == 0)
                fprintf(fp_out,
"%d, %12.6f, %12.6f, %12.6f, %12.6f, %12.6f, %12.6f, %12.6f, %12.6f, %12.6f¥n",
                        i, beta_[stage], gam_[stage], c_[stage], sgm_[stage],
                        logRProb_, logRProb_beta, logRProb_gam, logRProb_c, logRProb_sgm);
        }
        fprintf(fp_out, "¥n");

        if (stage >= 2)     // moderation of beta
            beta_[stage] = 0.1*beta_[stage-1]+0.9*beta_[stage];

        fprintf(fp_out, "result: ¥n");
        fprintf(fp_out, "beta_ =, %12.6f¥n", beta_[stage]);
        fprintf(fp_out, "gam_ =, %12.6f¥n", gam_[stage]);
        fprintf(fp_out, "c_ =, %12.6f¥n", c_[stage]);
        fprintf(fp_out, "sgm_ =, %12.6f¥n", sgm_[stage]);
        fprintf(fp_out, "logRProb_ =, %12.6f¥n", logRProb_);
        fprintf(fp_out, "¥n");

        beta_Rcd[stage] = beta_[stage];
        gam_Rcd[stage] = gam_[stage];
        c_Rcd[stage] = c_[stage];
        sgm_Rcd[stage] =sgm_[stage];
        logRProb_Rcd[stage] = logRProb_;

        // Prediction...Mountain Climbing, n: from NBgn to NEnd ... numerical differentiation
        fprintf(fp_out, "// Prediction...Mountain Climbing, n: from %d to %d ... numerical
differentiation¥n", NBgn, NEnd);
        fprintf(fp_out, "¥n");


        S[0] = S1[0];
        I[0] = I1[0];
        R[0] = R1[0];
        PrsI[0] = I[0];
        DlyI[0] = 0.0;
        AccI[0] = I[0];

        SIRsol(NBgn, NEnd, beta_[stage], gam_[stage]);
```

```
        if (mode == 1) {
            fprintf(fp_out, "n, t, S, I, R, PrsI, AccI, ObsPrsI1¥n");
            for(n = NBgn; n <= NEnd; n++) {
                if (n % nSkp == 0)
                    fprintf(fp_out,
"%d, %12.6f, %12.6f, %12.6f, %12.6f, %12.6f, %12.6f, %12.6f¥n",
                            n, (n+0.0)*dt, S[n], I[n], R[n], PrsI[n], AccI[n], ObsPrsI1[n]);
            }
        }
        else if (mode == 2) {
            fprintf(fp_out, "n, t, S, I, R, 10*DlyI, AccI, 10*ObsDlyI1¥n");
            for(n = NBgn; n <= NEnd; n++) {
                if (n % nSkp == 0)
                    fprintf(fp_out,
"%d, %12.6f, %12.6f, %12.6f, %12.6f, %12.6f, %12.6f, %12.6f¥n",
                            n, (n+0.0)*dt, S[n], I[n], R[n], 10.0*DlyI[n], AccI[n],
10.0*ObsDlyI1[n]);
            }
        }
        else if (mode == 3) {
            fprintf(fp_out, "n, t, S, I, R, 10*DlyI, AccI, ObsAccI1¥n");
            for(n = NBgn; n <= NEnd; n++) {
                if (n % nSkp == 0)
                    fprintf(fp_out,
"%d, %12.6f, %12.6f, %12.6f, %12.6f, %12.6f, %12.6f, %12.6f¥n",
                            n, (n+0.0)*dt, S[n], I[n], R[n], 10.0*DlyI[n], AccI[n],
ObsAccI1[n]);
            }
         }
        else
            fprintf(fp_out, "mode should be 1, 2 or 3¥n");
        fprintf(fp_out, "¥n");

    }


    // Prediction for whole period or beyond that...Mountain Climbing, n: from 1 to NE ...
numerical differentiation
```

```
    fprintf(fp_out, "// Prediction for whole period...Mountain Climbing, n: from 1 to %d ...
numerical differentiation¥n", NE);
    fprintf(fp_out, "¥n");


    fprintf(fp_out, "stage, beta_Rcd, gam_Rcd, c_Rcd, sgm_Rcd, logRProb_Rcd¥n");
    for (stage = 1; stage <= NStg; stage++)
        fprintf(fp_out, "%d, %12.6f, %12.6f, %12.6f, %12.6f, %12.6f¥n",
                stage, beta_Rcd[stage], gam_Rcd[stage], c_Rcd[stage], sgm_Rcd[stage],
logRProb_Rcd[stage]);
    fprintf(fp_out, "¥n");


    if (mode == 1) {
        fprintf(fp_out, "n, t, S, I, R, PrsI, AccI, ObsPrsI1¥n");
        for(n = 1; n <= NE; n++) {
            if (n % nSkp == 0)
                fprintf(fp_out,
"%d, %12.6f, %12.6f, %12.6f, %12.6f, %12.6f, %12.6f, %12.6f¥n",
                        n, (n+0.0)*dt, S[n], I[n], R[n], PrsI[n], AccI[n], ObsPrsI1[n]);
        }
    }
    else if (mode == 2) {
        fprintf(fp_out, "n, t, S, I, R, 10*DlyI, AccI, 10*ObsDlyI1¥n");
        for(n = 1; n <= NE; n++) {
            if (n % nSkp == 0)
                fprintf(fp_out,
"%d, %12.6f, %12.6f, %12.6f, %12.6f, %12.6f, %12.6f, %12.6f¥n",
                        n, (n+0.0)*dt, S[n], I[n], R[n], 10.0*DlyI[n], AccI[n],
10.0*ObsDlyI1[n]);
        }
    }
    else if (mode == 3) {
        fprintf(fp_out, "n, t, S, I, R, 10*DlyI, AccI, ObsAccI1¥n");
        for(n = 1; n <= NE; n++) {
            if (n % nSkp == 0)
                fprintf(fp_out,
"%d, %12.6f, %12.6f, %12.6f, %12.6f, %12.6f, %12.6f, %12.6f¥n",
```

```
                        n, (n+0.0)*dt, S[n], I[n], R[n], 10.0*DlyI[n], AccI[n],
ObsAccI1[n]);
        }
    }
    else
        fprintf(fp_out, "mode should be 1, 2 or 3¥n");
    fprintf(fp_out, "¥n");


    fclose(fp_inp);
    fclose(fp_out);

    pushKey();
}

/* ------------------------------------------------------------ */

void pushKey()
{
    printf("¥n      Push Return Key! ");
    getchar();
    getchar();
}

/* ------------------------------------------------------------ */

void func(double x, double y[], double f[])
{
    f[0] = y[1];
    f[1] = 1.0 - y[0] -2.0*y[1];
}

// ------------------------------------------------------------ //

double rand_normal( double myu, double sigma )
{
double z=sqrt( -2.0*log(Uniform()) ) * sin( 2.0*PI*Uniform() );
return myu + sigma*z;
 }
```

```
// ------------------------------------------------------------ //

double Uniform( void )
{
static int x=10;
int a=1103515245, b=12345, c=2147483647;
x = (a*x + b)&c;

return ((double)x+1.0) / ((double)c+2.0);
}

// ------------------------------------------------------------ //

double normal(double x, double myu, double sgm)              // 正規分布
{
    return 1.0/sqrt(2.0*PI*sgm*sgm)*exp(-(x-myu)*(x-myu)/2.0/sgm/sgm);
}

// ------------------------------------------------------------ //

double Pnml_I(int n, double c, double myu, double sgm)
{
    if (mode == 1)
        return normal((ObsPrsI1[n]-c*PrsI[n])/(c*PrsI[n]), myu, sgm);
    else if (mode == 2)
        return normal((ObsDlyI1[n]-c*DlyI[n])/(c*DlyI[n]), myu, sgm);
    else if (mode == 3)
        return normal((ObsAccI1[n]-c*AccI[n])/(c*AccI[n]), myu, sgm);
    else {
        fprintf(fp_out, "mode should be 1, 2 or 3¥n");
        return 1.0;
    }
}

// ------------------------------------------------------------ //

double logPnml_I(int n, double c, double myu, double sgm)
{
    return log(Pnml_I(n, c, myu, sgm));
```

```
}

// ------------------------------------------------------------------ //

void SIRsol(int N1, int N2, double beta, double gam)
{
    int n;

    for(n = N1; n <= N2; n++) {
        S[n] = S[n-1] - beta/(Npop+0.0)*S[n-1]*I[n-1]*dt;
        I[n] = I[n-1] + (beta/(Npop+0.0)*S[n-1]*I[n-1] - gam*I[n-1])*dt;
        R[n] = R[n-1] + gam*I[n-1]*dt;

        PrsI[n] = I[n];
        DlyI[n] = beta/(Npop+0.0)*S[n]*I[n];
        AccI[n] = AccI[n-1] + DlyI[n]*dt;
    }
}

// ------------------------------------------------------------------ //
```

(2) Input file: EstTimeSeqOSSMZ4TmDpdMtClmbChgPrmYFineTime_inp.dat

```
Case            4
mode            1
Npop            32000
beta1           0.2
gam1            0.05
c1              0.7
S1[0]           31998
I1[0]           2
R1[0]           0
sgmObsAccI1     0.025
sgmObsAccI2     0.25
betacon         1
gamcon          0
ccon            0
sgmcon          0
betaini         0.16
```

gamini	0.04
cini	0.56
sgmini	0.2
NS	3600
NE	3600
NStg	36
WStg	100
dt	0.1
dlmd	0.00000001
lmd	0.0000002
iEnd	25000
iSkp	250
nSkp	10
oSkp	10
PrtCtrl	0

7. DIAGNOSIS OF ILLNESS BASED ON CHECKLIST

In the future, medical diagnosis by artificial intelligence is expected to become widespread. Therefore, we considered a method for diagnosing a disease based on a checklist using Bayesian inference. In this chapter, we describe the development of basic theory and algorithms for that purpose. Bayesian inference can provide learning and inference that is completely different from neural networks.

Although it has been widely believed that neural networks, which are said to imitate the structure of the human brain, have great potential, they faced the major problem of increasing the scale of networks and making learning difficult. Therefore, it could only learn childish problems. Against this background, deep learning [1,2,3] developed by Hinton et al appeared. As a result, a breakthrough has occurred in technologies such as image recognition, speech recognition, and machine translation that humans can easily do but machines cannot.

With the birth of deep learning, it has become possible to increase the scale and accuracy of neural networks, and the learning ability has dramatically improved, enabling advanced learning. Image recognition, voice recognition, and machine translation, which were thought to be almost impossible, have become possible. Nowadays, neural networks or deep learning have become synonymous with artificial intelligence (AI), and soon there is even a view that all human intellectual work will be possible with AI.

However, there are problems such as "inference is a black box", "unexpected answer due to overfitting", and "large-scale network and long-time learning". Isn't deep learning far from perfect? It is hoped that these problems will be overcome as soon as possible. Above all, the black-box nature of inference is a fundamental problem.

Bayesian inference is based on learning and inference that is completely different from neural networks [4,5,6]. It may be unrelated to some problems of neural networks. The Bayesian inference seems to have disappeared in the world of statistics for a while. This is because prior probabilities have a strong subjective aspect and have been refused by some orthodox statisticians. However, when it comes to decision-making on business issues, it would not be possible to make decisions that are based on objective assumptions in every sense. In such a case, Bayesian inference that can incorporate subjectivity will be extremely practical.

In neural network learning, the difference between the neuron value of the output layer and the teacher data for a certain input is regarded as an error, and the weight is adjusted so that the error is minimized. That is, we have to solve the multivariable minimum value problem of a large scale. On the other hand, in Bayesian learning using Bayesian inference, the learning refers to finding the probability distribution given as a likelihood function from a large number of training data. In Bayesian learning, it is considered that learning is to find the frequency distribution from the learning data. Therefore, Bayesian learning is free from the various difficulties associated with the multivariable minimum value problem.

On the other hand, in Bayesian inference, how to set prior probabilities can be a problem. If there is only one source of data, setting prior probabilities does not matter. If the number of data is increased, it converges to one answer regardless of the setting of prior probabilities. For example, this is the case with the problem of parameter estimation of the probability distribution discussed in chapter 3. This can be shown either numerically or theoretically using conjugate prior probabilities.

However, if there are as many data sources as there are symptoms as discussed in this chapter, the posterior probabilities will change depending on the prior probability settings. Therefore, in such a case, it is preferable that the prior probability can be set based on the actual value. On the other hand, in decision-making problems such as management decisions, there may be a limit to the objective setting of prior probabilities. In such a case, when solving the exact same problem, the result will change depending on the person, but this is unavoidable. It should be considered that the feature of Bayesian inference is to be able to show some guidelines even in such a case.

In this chapter, we discuss a method of performing disease diagnosis using checklist data by Bayesian inference as an example of disease diagnosis by artificial intelligence, which is expected to increase in the future. As such a disease diagnosis, in addition to the data of the presence or absence of fever (0/1) according to the checklist, the data of measured values such as body temperature and blood pressure, or a mixture of these may be used. The checklist method is considered to be the simplest.

7.1. Diagonosis based on checklist

Suppose that the conditional probability $P\,(Result \mid Cause)$ of the *Result* under a certain *Cause* is known. Bayes' theorem is to find the inverse probability, that is, the inverse probability $P\,(Cause \mid Result)$ of the *Cause* that caused such a *Result*. Bayesian inference uses Bayes' theorem to estimate the *Cause* for a *Result* based on pre-estimated results.

Bayes' theorem is given by

$$P(Cause \mid Result) = \frac{P(Result, Cause)}{P(Result)} = \frac{P(Result \mid Cause)P(Cause)}{P(Result)}. \tag{7.1}$$

This is just a mathematical theorem. A concrete example shows how this theorem is used to estimate the cause.

In the present paper, Bayesian inference is used to distinguish among the new coronavirus infection (COVID-19), influenza, colds, and allergies. The state of Minnesota in the United States details the pandemic of COVID-19, and the figure shown in Fig. 7.1 is posted in it. Based on this figure, the above judgment is made (https://en.wikipedia.org/wiki/COVID-19_pandemic_in_Minnesota).

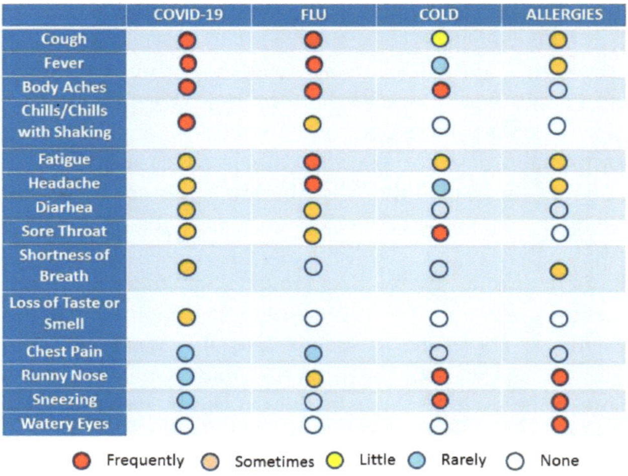

Fig. 7.1. COVID-19 Symptoms vs. Flu, Cold & Allergies
(https://www.co.carver.mn.us/home/showdocument?id=19659).

First, the frequency of symptoms shown by the red, brown, yellow, blue, and white circles shown in Fig. 7.1 is quantifized as shown in Table 1.

Table 7.1. Degree of Quantified Symptoms.

Color of circle	**Explanation**	**Probability**	
		With symptom	**Without symptom**
Red	Frequently	0.95	0.05
Brown	Sometimes	0.5	0.5
Yellow	Little	0.25	0.75
Blue	Rarely	0.1	0.9
White	None	0.05	0.95

When a doctor examines a patient, a 14-dimensional vector $\mathbf{S}=(S_1, S_2, S_3, S_4, S_5, S_6, S_7, S_8, S_9, S_{10}, S_{11}, S_{12}, S_{13}, S_{14})$ is obtained. S_i takes a binary value of 1 or 0. The value 1 or 0 represents "with" or "without" of the symptom S_i of the patient's disease D, respectively. In the following, this is called a symptom checklist.

When using Bayesian inference, we introduce the probability and think as follows. Each symptom n (= 1,2,…, 14) of infectious disease D_i (i = 1,2,…, 4) is represented by a stochastic variable S_{in} that takes a binary value of 1 with symptom and 0 without symptom. Let the probability be $P(S_{in}|D_i)$.

The probability $P(S_{in}|D_i)$ of symptom S_{in} in infectious disease D_i is given in Fig. 7.1 and Table 7.1. Symptom S_{in} is a binary variable, with S_{in} = 1 with symptoms and $S_{in} = 0$ without symptoms. For example, in COVID-19, according to with or without cough, the probability P (S_{11} = 1, 0 | D_1) is given by 0.95, 0.05. Suppose that a patient's illness D is examined and the symptom checklist is S = (S_1, S_2,…,S_{14}). The inverse probability P (D_i | S) for estimating whether this patient has infectious diseases D_i, i = 1,2, ..., 4 is given by Bayes' theorem as follows:

$$P(\mathrm{D}_i \mid \mathbf{S}) = \frac{P(\mathbf{S},\mathrm{D}_i)}{P(\mathbf{S})} = \frac{P(\mathbf{S} \mid \mathrm{D}_i)P(\mathrm{D}_i)}{P(\mathbf{S})} = \frac{P(\mathbf{S} \mid \mathrm{D}_i)P(\mathrm{D}_i)}{\sum_{j=1}^{4} P(\mathbf{S} \mid \mathrm{D}_j)P(\mathrm{D}_j)}. \tag{7.2}$$

If we assume for the prior provability as

$$P(\mathrm{D}_1) = P(\mathrm{D}_2) = \cdots = P(\mathrm{D}_4), \tag{7.3}$$

we obtain from equation (2)

$$P(\mathrm{D}_i \mid \mathbf{S}) = \frac{P(\mathbf{S} \mid \mathrm{D}_i)}{\sum_{j=1}^{4} P(\mathbf{S} \mid \mathrm{D}_j)}. \tag{7.4}$$

This is nothing but the maximum likelihood method. In addition, P(**S**|D_i) is calculated as follows:

$$P(\mathbf{S} \mid \mathrm{D}_i) = P(S_1 \mid \mathrm{D}_i)P(S_2 \mid \mathrm{D}_i)\cdots P(S_n \mid \mathrm{D}_i). \tag{7.5}$$

Concerning the prior probabilities, we have introduced an assumption such as given by equation (7.3). In the case of one data source such as parameter estimation of the probability distribution discussed in chapter 3, if the number of data increases, the result does not depend on how to obtain prior probabilities. However, in the case of this example, since the data, that is, each symptom, is a source of different data, the estimation result may differ depending on the setting of the prior probability. This will be described below.

An example of numerical calculation based on the above theory is shown below. First, a patient symptom checklist **S** = (S_1, S_2, …, S_{14}) is generated to satisfy the probabilities given in Fig. 7.1 and Table 7.1. **S** is generated virtually using a function that generates a random number uniformly distributed in [0, 1]. For example, if the infectious disease of the patient is D_i (for example, COVID-19), the probability that the *n*-th symptom S_{in} is positive is P (S_{in} | D_i), and the following is used to determine the value of S_n:

$$S_n = \begin{cases} 1, \text{if } \mathit{uniform}() \le P(S_{in} \mid D_i) \\ 0, \text{otherwise} \end{cases}, \tag{7.6}$$

where $\mathit{uniform}()$ is a function generating a uniform random number in [0,1].

Instead of the actual patient checklist, the following calculation uses equation (7.6) to generate a virtual checklist, which is used as input. For example, using the following checklist generated by equation (7.6) for patients with COVID-19:

$$\mathbf{S} = (1, 1, 1, 1, 0, 1, 0, 0, 0, 0, 0, 0, 0, 1), \tag{7.7}$$

into equation (7.4), we obtain

$$P(\mathbf{S} \mid \mathrm{D}_1) = 0.825,\ P(\mathbf{S} \mid \mathrm{D}_2) = 0.175,\ P(\mathbf{S} \mid \mathrm{D}_3) = 0.0000005,\ P(\mathbf{S} \mid \mathrm{D}_4) = 0.0003. \tag{7.8}$$

Therefore, this patient is presumed to have COVID-19, which has the highest probability.

Similarly, if we substitute a checklist for an influenza patient generated by equation (7.6):

$$\mathbf{S} = (1, 1, 1, 1, 1, 1, 0, 0, 0, 0, 0, 1, 0, 1), \tag{7.9}$$

into equation (7.4), we obtain

$$P(\mathbf{S} \mid \mathrm{D}_1) = 0.027,\ P(\mathbf{S} \mid \mathrm{D}_2) = 0.972,\ P(\mathbf{S} \mid \mathrm{D}_3) = 0.000003,\ P(\mathbf{S} \mid \mathrm{D}_4) = 0.001. \tag{7.10}$$

Therefore, the patient is presumed to have influenza.

Similarly, if we substitute a checklist for a cold patient and that for an allergy patient generated by equation (7.6):

$$\mathbf{S} = (1, 0, 1, 0, 0, 1, 0, 1, 0, 0, 0, 1, 1, 1) \tag{7.11}$$

and

$$\mathbf{S} = (1, 0, 0, 0, 0, 1, 0, 0, 0, 0, 0, 1, 1, 1) \tag{7.12}$$

into equation (7.4), we obtain

$$P(\mathbf{S} \mid \mathrm{D}_1) = 0.00004,\ P(\mathbf{S} \mid \mathrm{D}_2) = 0.0007\ P(\mathbf{S} \mid \mathrm{D}_3) = 0.866,\ P(\mathbf{S} \mid \mathrm{D}_4) = 0.133 \tag{7.13}$$

and

$$\begin{aligned} &P(\mathbf{S} \mid \mathrm{D}_1) = 0.00000004,\ P(\mathbf{S} \mid \mathrm{D}_2) = 0.0000007, \\ &P(\mathbf{S} \mid \mathrm{D}_3) = 0.00005,\ P(\mathbf{S} \mid \mathrm{D}_4) = 1.000, \end{aligned} \tag{7.14}$$

respectively. Therefore, the patients are presumed to have cold and allergy, respectively.

Let's consider the effects of the prior probabilities. Equation (7.4) is obtained assuming that all prior probabilities are the same. If the prior probabilities are different, there is a possibility that the inference results might change. For example, suppose the prior probabilities are assumed as

$$
\begin{aligned}
&P(D_1)=0.25,\ P(D_2)=0.25,\ P(D_3)=0.25,\ P(D_4)=0.25\\
&P(D_1)=0.10,\ P(D_2)=0.20,\ P(D_3)=0.30,\ P(D_4)=0.40\\
&P(D_1)=0.40,\ P(D_2)=0.30,\ P(D_3)=0.20,\ P(D_4)=0.10,\\
&P(D_1)=0.10,\ P(D_2)=0.40,\ P(D_3)=0.40,\ P(D_4)=0.10\\
&P(D_1)=0.40,\ P(D_2)=0.10,\ P(D_3)=0.10,\ P(D_4)=0.40
\end{aligned}
\tag{7.15}
$$

then, the same results are obtained by equation (7.2) for the inputs of equations (7.7), (7.9), (7.11), and (7.12).

However, if we assume the following values for the prior probabilities:

$$
\begin{aligned}
&P(D_1)=0.01,\ P(D_2)=0.49,\ P(D_3)=0.49,\ P(D_4)=0.01,\\
&P(D_1)=0.49,\ P(D_2)=0.01,\ P(D_3)=0.01,\ P(D_4)=0.49,
\end{aligned}
\tag{7.16a,b}
$$

we have different results by equation (7.2) as shown in Table 7.2. The input 1, 2, 3, and 4 mean equations (7.7), (7.9), (7.11), and (7.12). The maximum of the probability gives the diagnosis of the disease. The diagnosis on the non-diagonal means the result is not correct.

Table 7.2a Inference Results
(Using Prior Probability Given by equation (7.16a)

	i = 1	i = 2	i = 3	i = 4
$P(D_1,\mathbf{S}i)$	0.088	10^{-4}	10^{-7}	10^{-8}
$P(D_2,\mathbf{S}i)$	0.912	0.999	10^{-4}	10^{-5}
$P(D_3,\mathbf{S}i)$	10^{-6}	10^{-6}	0.996	0.002
$P(D_4,\mathbf{S}i)$	10^{-5}	10^{-5}	0.003	0.998

Table 7.2b Inference Results
(Using Prior Probability Given by equation (7.16b)

	i = 1	i = 2	i = 3	i = 4
$P(D_1,\mathbf{S}i)$	0.995	0.557	10^{-4}	10^{-8}
$P(D_2,\mathbf{S}i)$	0.004	0.412	10^{-5}	10^{-8}
$P(D_3,\mathbf{S}i)$	10^{-8}	10^{-6}	0.117	10^{-8}
$P(D_4,\mathbf{S}i)$	10^{-4}	0.031	0.883	1.0

One way for treating this kind of situation would be to adjust the prior probabilities so that the order of the prior probabilities coincides with that of posterior probabilities since the prior probabilities based on an objective basis could reflect the posterior probabilities. The result of posterior probabilities due to the prior probabilities given by equation (7.16a) is shown in Fig. 7.2a. The probability of disease means the normalized probability:

$$P(D_i)\sum_{n'=1}^{n} P(S_n \mid D_i) \Bigg/ \left[\sum_{i=1}^{4} P(D_i) \sum_{n'=1}^{14} P(S_n \mid D_i) \right] \tag{7.17}$$

The order of D_1 and D_3 is different between the prior and posterior probabilities.

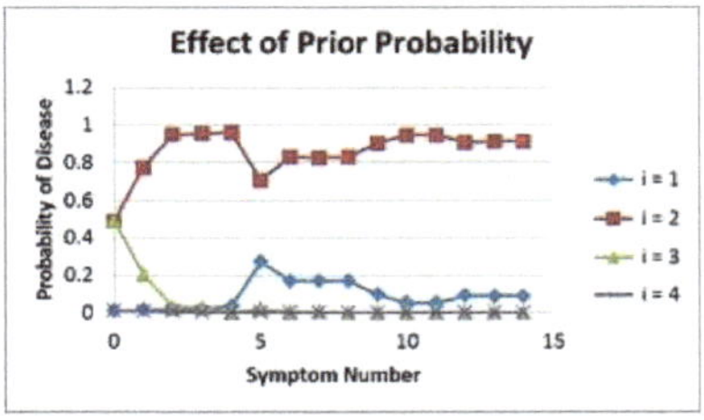

Fig. 7.2a. Prior Probability: $P(D_1)=0.01$, $P(D_2)=0.49$, $P(D_3)=0.49$, $P(D_4)=0.01$.

If we adjust the prior probabilities, we obtain the result as shown in Fig. 7.2b. The order of D_1 and D_2 is different between the prior and posterior probabilities.

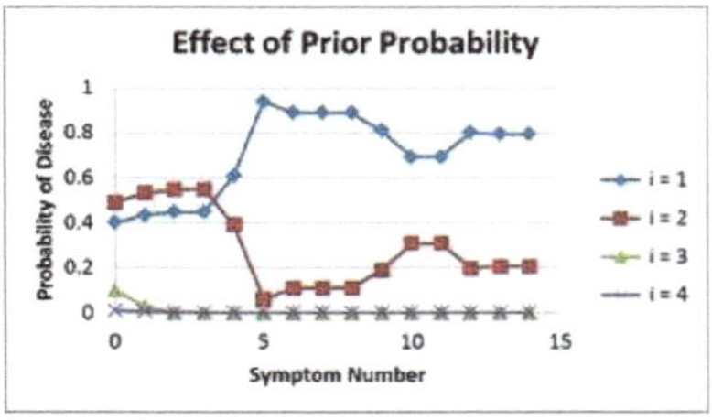

Fig. 7.2b. Prior Probability: $P(D_1)=0.40$, $P(D_2)=0.49$, $P(D_3)=0.10$, $P(D_4)=0.01$.

If we adjust the prior probabilities, we obtain the result as shown in Fig. 7.2c. The order of D_3 and D_4 is different between the prior and posterior probabilities.

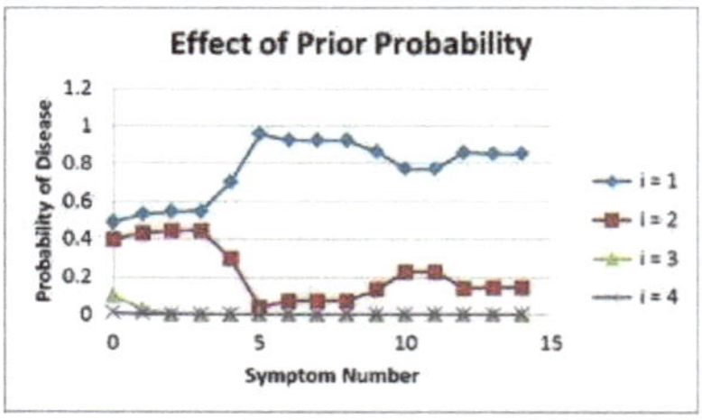

Fig. 7.2c. Prior Probability: $P(D_1)=0.49$, $P(D_2)=0.40$, $P(D_3)=0.10$, $P(D_4)=0.01$.

If we adjust the prior probabilities, we obtain the result as shown in Fig. 7.2d. The order seems appropriate.

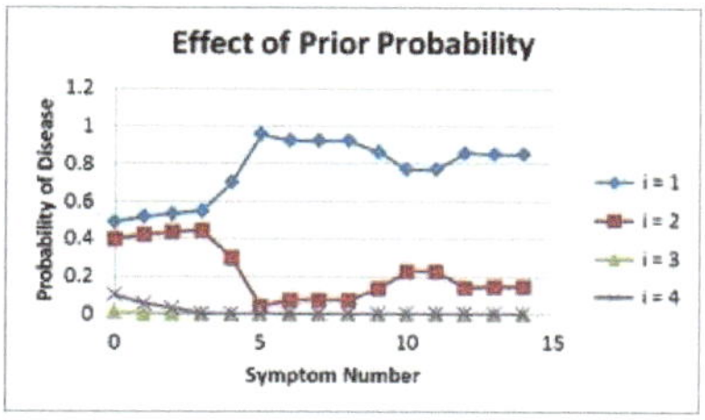

Fig. 7.2d. Prior Probability: $P(D_1)=0.49$, $P(D_2)=0.40$, $P(D_3)=0.01$, $P(D_4)=0.1$.

To investigate the inference accuracy, we calculated the posterior probabilities 1,000 times for each of the inputs generated probabilistically by equation (7.6) for each of the diseases, where the prior probabilities are assumed to be equal. The results are shown in Table 7.3. It could be considered that the values on the diagonal show the correct inference. The accuracy of more than 90% is obtained.

Table 7.3 Inference Results

	D_1	D_2	D_3	D_4
Diagnosis 1	90.7%	4.9%	0%	0%
Diagnosis 2	8.9%	94.5%	0.7%	0.2%
Diagnosis 3	0.4%	0.5%	98.9%	0.4%
Diagnosis 4	0%	0.1%	0.4%	99.4%

If the accuracy of inference is poor, we should reconsider the frequencies of symptoms and/or add the additional symptoms.

The programing code is shown in Appendix 7A.

Recently, various applications using deep learning are widely applied. However, it couldn't be almighty. It takes a long time to study, and it outputs abnormal answer because of overfitting. Since the biggest problem would be the black box nature and the basis of the reasoning is unclear, it makes us anxious.

Since Bayesian inference is based on the Bayesian theorem in statistics, it is based on a completely different theory. The method of learning is also completely different. Hence, it has a possibility of liberating us from the above-mentioned worries.

The present report only shows a basic idea of how to make the diagnosis based on a checklist using Bayesian inference. We must wait for future studies to clarify what kind of the possibility it may have.

Furthermore, we made some investigations on the prior probability that always becomes controversial, when we use Bayesian inference.

REFERENCES 7

[1] Krizhevsky, A., Sutskever, I. & Hinton, G. ImageNet classification with deep convolutional neural networks. In Proc. Advances in Neural Information Processing Systems 25 1090–1098 (2012).

[2] Yann LeCun1, Yoshua Bengio & Geoffrey Hinton, Deep learning, NATURE | VOL 521 | 28 MAY (2015).

[3] Masato Taki, Introduction to Deep Learning, Kodansha (2017) in Japanese.

[4] Nozomu Matsubara, Introduction to Bayesian Statistics, Tokyo Tosho (2008) in Japanese.

[5] Atsushi Suyama, Introduction to Mchine Learning by Bayesian Inference, Kodansha (2017) in Japanese.

[6] Hiroshi Isshiki, Pattern Recognition by Bayesian Inference, The 63rd Joint Meeting of Automatic Control (2020), https://www.jstage.jst.go.jp/article/jacc/63/0/63_329/_pdf.

Appendix 7A Code for diagnosis of illness based on checklist

"Microsoft C/C++ Compiler Version 17.00.50727.1 for x86" and "Microsoft Linker Version 11.00.50727" were used for compile and link (in command window; cl source_file_name.c).

(1) Programing code: MedicalDiagnosisCheckSheetXPriorX.c

```
// ---------------------------------------------------------------------- //
//                                                                        //
// File Name: ProbDistMyuZNewX.c                  2020.12.13-2020.12.14 //
//                                                                        //
// File Name: MedicalDiagnosisUniform.c           2021.01.06-2021.01.07 //
// File Name: MedicalDiagnosisUniformX.c          2021.01.07-2021.01.07 //
//                                                                        //
// File Name: MedicalDiagnosisCheckSheet.c        2021.01.08-2021.01.08 //
// File Name: MedicalDiagnosisCheckSheetX.c       2021.01.08-2021.01.08 //
// File Name: MedicalDiagnosisCheckSheetXPrior.c  2021.01.08-2021.01.08 //
// File Name: MedicalDiagnosisCheckSheetXPriorX.c 2021.01.08-2021.01.09 //
// File Name: MedicalDiagnosisCheckSheetXPriorX.c 2022.05.24-2022.05.24 //
//                                                                        //
//    Medical DiagnosisX by Bayes Inference                               //
//                                                                        //
// ---------------------------------------------------------------------- //

// ----- functions ------------------------------------------------------ //
```

```
#include <stdio.h>
#include <stdlib.h>
#include <string.h>
#include <math.h>

#define PI      3.14159265     // Pi

void main();
void pushKey();

double Uniform( void );                            // uniform random number
double rand_normal(double,double);                 // normal random number

double normal(double,double,double);               // normal distribution

void probability(void);                            // setting of probability

int imax_RProbNrm();                               // max value

// ----- variables ------------------------------------------------------- //

char title_memo[5000];

int I;                          // number of disease ... 4
int N;                          // number of symptom ... 14

double Red;                     // degree of frequency
double Brown;                   // degree of frequency
double Yellow;                  // degree of frequency
double Blue;                    // degree of frequency
double White;                   // degree of frequency

double Prob[5][15];             // probability symptom positive

int S1[15];                     // symptom of a patient ... 1/0: positive/negative

double RProb[5];                // reverse probability
double RProbNrm[5];             // normalized probability
```

```
double RProbNrm1[15][5];          // normalized probability

double PriProb[5];                // prior probability

int DNo;                          // disease number

double AMAT[1001][2001];          // matrix

FILE *fp_inp;                     // file pointer of input file
FILE *fp_out;                     // file pointer of output file

char InputDataFile[80];           // input file name
char OutputDataFile[80];          // output file name

char buf[5000];

int Num;                          // number of data
double Value[10001];              // value
double histo[10001];              // frequency
double x_max;
double x_min;
double Dx;
double width;

// ------------------------------------------------------------ //

void probability(void)
{
    Prob[1][1]  = Red;
    Prob[1][2]  = Red;
    Prob[1][3]  = Red;
    Prob[1][4]  = Red;
    Prob[1][5]  = Brown;
    Prob[1][6]  = Brown;
    Prob[1][7]  = Brown;
    Prob[1][8]  = Brown;
    Prob[1][9]  = Brown;
    Prob[1][10] = Brown;
```

```
Prob[1][11] = Blue;
Prob[1][12] = Blue;
Prob[1][13] = Blue;
Prob[1][14] = White;

Prob[2][1]  = Red;
Prob[2][2]  = Red;
Prob[2][3]  = Red;
Prob[2][4]  = Brown;
Prob[2][5]  = Red;
Prob[2][6]  = Red;
Prob[2][7]  = Brown;
Prob[2][8]  = Brown;
Prob[2][9]  = White;
Prob[2][10] = White;
Prob[2][11] = Blue;
Prob[2][12] = Brown;
Prob[2][13] = White;
Prob[2][14] = White;

Prob[3][1]  = Yellow;
Prob[3][2]  = Blue;
Prob[3][3]  = Red;
Prob[3][4]  = White;
Prob[3][5]  = Brown;
Prob[3][6]  = Blue;
Prob[3][7]  = White;
Prob[3][8]  = Red;
Prob[3][9]  = White;
Prob[3][10] = White;
Prob[3][11] = White;
Prob[3][12] = Red;
Prob[3][13] = Red;
Prob[3][14] = White;

Prob[4][1]  = Brown;
Prob[4][2]  = Brown;
Prob[4][3]  = White;
Prob[4][4]  = White;
```

```
 Prob[4][5]  = Brown;
 Prob[4][6]  = Brown;
 Prob[4][7]  = White;
 Prob[4][8]  = White;
 Prob[4][9]  = Brown;
 Prob[4][10] = White;
 Prob[4][11] = White;
 Prob[4][12] = Red;
 Prob[4][13] = Red;
 Prob[4][14] = Red;
}

// ------------------------------------------------------------ //

int imax_RProbNrm()              // maximum
{
    int i, i_max;
    double tmp;

    i_max = 1;
    tmp = RProbNrm[1];
    for (i = 2; i <= 4; i++)
        if (tmp <= RProbNrm[i]) {
            tmp = RProbNrm[i];
            i_max = i;
        }

    return i_max;
}

// ------------------------------------------------------------ //

void main()
{
    int i, j, n, ntmp;
    double sum, tmp;


    //// open input file
```

```
sprintf(InputDataFile, "MedicalDiagnosisCheckSheetXPriorX_inp.dat");

if ((fp_inp = fopen(InputDataFile, "r")) == NULL) {
    printf("Failed in Reading Input Data File! ... %s¥n", InputDataFile);
    exit(1);
}

//// open output file
sprintf(OutputDataFile, "MedicalDiagnosisCheckSheetXPriorX_out.csv");

if ((fp_out = fopen(OutputDataFile, "w")) == NULL) {
    printf("Failed in Reading Output Data File! ... %s¥n", OutputDataFile);
    exit(1);
}


//// input from file
fscanf(fp_inp, "%s", title_memo);

 fscanf(fp_inp, "%s %lf", buf, &Red);
 fscanf(fp_inp, "%s %lf", buf, &Brown);
 fscanf(fp_inp, "%s %lf", buf, &Yellow);
 fscanf(fp_inp, "%s %lf", buf, &Blue);
 fscanf(fp_inp, "%s %lf", buf, &White);

 fscanf(fp_inp, "%s %lf", buf, &PriProb[1]);
 fscanf(fp_inp, "%s %lf", buf, &PriProb[2]);
 fscanf(fp_inp, "%s %lf", buf, &PriProb[3]);
 fscanf(fp_inp, "%s %lf", buf, &PriProb[4]);


 printf("memo: %s¥n", title_memo);
 printf("¥n");

 printf("Red    = %12.6f¥n", Red);
 printf("Brown  = %12.6f¥n", Brown);
 printf("Yellow = %12.6f¥n", Yellow);
 printf("Blue   = %12.6f¥n", Blue);
 printf("White  = %12.6f¥n", White);
```

```
printf("¥n");

printf("PriProb[1] = %12.6f¥n", PriProb[1]);
printf("PriProb[2] = %12.6f¥n", PriProb[2]);
printf("PriProb[3] = %12.6f¥n", PriProb[3]);
printf("PriProb[4] = %12.6f¥n", PriProb[4]);
printf("¥n");

fprintf(fp_out, "memo: %s¥n", title_memo);
fprintf(fp_out, "¥n");

fprintf(fp_out, "Red    = %12.6f¥n", Red);
fprintf(fp_out, "Brown  = %12.6f¥n", Brown);
fprintf(fp_out, "Yellow = %12.6f¥n", Yellow);
fprintf(fp_out, "Blue   = %12.6f¥n", Blue);
fprintf(fp_out, "White  = %12.6f¥n", White);
fprintf(fp_out, "¥n");

fprintf(fp_out, "PriProb[1] = %12.6f¥n", PriProb[1]);
fprintf(fp_out, "PriProb[2] = %12.6f¥n", PriProb[2]);
fprintf(fp_out, "PriProb[3] = %12.6f¥n", PriProb[3]);
fprintf(fp_out, "PriProb[4] = %12.6f¥n", PriProb[4]);
fprintf(fp_out, "¥n");

fprintf(fp_out, "Red    = %12.6f¥n", Red);
fprintf(fp_out, "Brown  = %12.6f¥n", Brown);
fprintf(fp_out, "Yellow = %12.6f¥n", Yellow);
fprintf(fp_out, "Blue   = %12.6f¥n", Blue);
fprintf(fp_out, "White  = %12.6f¥n", White);
fprintf(fp_out, "¥n");

pushKey();


//// probability of symptom positive
probability();

fprintf(fp_out, "probability of positive and negative¥n");
fprintf(fp_out, "i=1:, COVID-19, i=2:, Flu, i=3:, Cold, i=4:, Allergies¥n");
```

```
for (i = 1; i <= 4; i++) {
    fprintf(fp_out, "i = %d¥n", i);
    fprintf(fp_out, "n, Prob[%d], 1.0-Prob[%d]¥n", i, i);
    for (n = 1; n <= 14; n++)
        fprintf(fp_out, "%d, %12.6f, %12.6f¥n", n, Prob[i][n], 1.0-Prob[i][n]);
    fprintf(fp_out, "¥n");
}
//// input symptom
fprintf(fp_out, "judge the disease¥n");
printf("¥n");
printf("Input Deseace Number ? ... 1-4:  ");
scanf("%d", &DNo);

for (n = 1; n <= 14; n++) {
    if (Uniform() <= Prob[DNo][n])
        S1[n] = 1;
    else
        S1[n] = 0;
}

printf("diagnosis:, DNo = %d¥n", DNo);
for (n = 1; n <= 14; n++)
    printf("n1 = %2d, S1 = %d¥n", n, S1[n]);
printf("¥n");

fprintf(fp_out, "diagnosis:, DNo =, %d¥n", DNo);
fprintf(fp_out, "n, S1¥n");
for (n = 1; n <= 14; n++)
    fprintf(fp_out, "%d, %d¥n", n, S1[n]);
fprintf(fp_out, "¥n");
//// make a diagnosis
for (ntmp = 1; ntmp <= 14; ntmp++) {
    for (i = 1; i <= 4; i++) {
        RProb[i] = PriProb[i];
        for (n = 1; n <= ntmp; n++)
            if (S1[n] == 1)
                RProb[i] *= Prob[i][n];
            else
                RProb[i] *= 1.0-Prob[i][n];
```

```
    }

    sum = 0.0;
    for (i = 1; i <= 4; i++)
        sum += RProb[i];

    for (i = 1; i <= 4; i++)
        RProbNrm[i] = RProb[i]/sum;

    fprintf(fp_out, "ntmp =, %d, DNo =, %d¥n", ntmp, DNo);
    fprintf(fp_out, "i, RProb, RProbNrm¥n");
    for (i = 1; i <= 4; i++)
        fprintf(fp_out, "%d, %12.6g, %12.6g¥n", i, RProb[i], RProbNrm[i]);
    fprintf(fp_out, "¥n");

    for (i = 1; i <= 4; i++)
        RProbNrm1[ntmp][i] = RProbNrm[i];

    DNo = imax_RProbNrm();
    printf("ntmp = %d, Disease is %d.¥n", ntmp, DNo);
    fprintf(fp_out, "ntmp =, %d, Disease is, %d¥n", ntmp, DNo);
    fprintf(fp_out, "¥n");
}


fprintf(fp_out, "ntmp, i = 1, i = 2, i = 3, i = 4¥n");
fprintf(fp_out, "%d, %12.6g, %12.6g, %12.6g, %12.6g¥n",
        0, PriProb[1], PriProb[2], PriProb[3], PriProb[4]);
for (ntmp = 1; ntmp <= 14; ntmp++) {
    fprintf(fp_out, "%d, ", ntmp);
    for (i = 1; i <= 4; i++)
        fprintf(fp_out, "%12.6g, ", RProbNrm1[ntmp][i]);
    fprintf(fp_out, "¥n");
}
fprintf(fp_out, "¥n");


fclose(fp_out);
```

```
    pushKey();
}

// ------------------------------------------------------------ //

void pushKey()
{
    printf("¥n       Push Return Key! ");
    getchar();
    getchar();
}

// ------------------------------------------------------------ //

double rand_normal( double myu, double sigma )
{
//      double z=sqrt( -2.0*log(Uniform()) ) * sin( 2.0*M_PI*Uniform() );
double z=sqrt( -2.0*log(Uniform()) ) * sin( 2.0*PI*Uniform() );
return myu + sigma*z;
 }

// ------------------------------------------------------------ //

double Uniform( void )    // [0,1] random number
{
static int x=10;
int a=1103515245,b=12345,c=2147483647;
x = (a*x + b)&c;

return ((double)x+1.0) / ((double)c+2.0);
}

// ------------------------------------------------------------ //

double normal(double x, double myu, double sgm)        // normal distribution
{
    return 1.0/sqrt(2.0*PI*sgm*sgm)*exp(-(x-myu)*(x-myu)/2.0/sgm/sgm);
}
```

// -- //

(2) Input data: MedicalDiagnosisCheckSheetXPriorX_inp.dat

```
ProbDistParam_20220524
Red                     0.95
Brown                   0.5
Yellow                  0.25
Blue                    0.1
White                   0.05
PriProb[1]              0.01
PriProb[2]              0.49
PriProb[3]              0.49
PriProb[4]              0.01
```

8. PATTERN RECOGNITION

Deep learning has brought about a big breakthrough in technologies such as image recognition, speech recognition, and machine translation that humans can easily do but machines cannot. Now that neural networks or deep learning have become synonymous with artificial intelligence (AI), the idea that all of the human intellectual work in not far future is possible with AI has come to be born. But deep learning is just one part of AI. It is also known that there are serious problems such as "inference is a black box". Bayesian inference can provide learning and inference that is completely different from neural networks. Therefore, it may be possible to overcome the problems of neural networks. In this paper, we discuss the application of Bayesian inference to pattern recognition of MNIST hand-written numbers [1,2,9].

8.1. Bayesian inference and patter recognition

What is human memory? Neural networks are said to imitate the human brain [3,4,5]. The brain has an enormous number of brain cells, which are said to be in the hundreds of billions, and it is said that the brain cells are connected to each other by nerve fibers called axons to form a network. This network has the function of transmitting and controlling information between brain cells, and what kind of information processing is performed is formed by learning. Neural networks are mathematical models of this hypothesis. Therefore, the memory is stored in the network.

There is another way of thinking. It is called the grandmother hypothesis, and it is a hypothesis that there is a cell that reacts specifically when you look at the grandmother. This means that memory is stored in a single cell or multiple cells, which is very different from the network theory described above. For example, the handwritten character "1" is formed in brain cells by looking at various "1" s. The frequency distribution (probability distribution) is made in a grid pattern for the character "1". It can be considered that the grid pattern for the character "1" has become established in the cell set in this way. This kind of thinking arises from the mathematical modeling of numbers pattern recognition by Bayesian inference.

In this way, Bayesian inference is based on learning and inference that are completely different from neural networks [6,7,8,9]. Bayesian inference may be unrelated to some of the problems with neural networks.

The Bayesian inference seems to have disappeared in the world of statistics for a while. This is because prior probabilities have a strong subjective aspect and have been repelled by some orthodox statisticians. However, when it comes to decision-making on issues such as management, politics, and society, it seems that judgments based on objective assumptions in every sense cannot be hoped for. In such a case, Bayesian inference that can incorporate subjectivity will be extremely practical.

In the training of the neural network, the difference between the value of the neuron in the output layer and the teacher data for a certain input is regarded as an error, and the weight is adjusted so that the error is minimized. That is, we have to solve the multi-variable minimum value problem of a vast number of unknown variables. On the other hand, in Bayesian learning using Bayesian inference, it means to obtain the probability distribution of the values to be taken by the cells constituting the pattern from a large number of learning data. In Bayesian learning, learning is to find the frequency distribution from the learning data. Therefore, it is free from the various difficulties associated with the multi-variables minimum value problem.

However, in the case of Bayesian inference, the problem of the maximum value of multiple variables appears also at the judgment stage for finding the maximum probability. Generally speaking, it is not a huge number of variables such as network weights encountered in the learning stage of neural networks, but a maximum value problem of a relatively small number of variables.

Suppose you know the conditional probability P (Result | Cause) of the Result under a Cause. Bayes' theorem is to find the inverse probability, that is, the inverse probability P (Cause | Result) of the cause that caused such a result based on the result estimated in advance. The Bayesian inference estimates the cause of the problem.

The Bayes' theorem is given by

$$P(Cause\,|\,Result) = \frac{P(Result, Cause)}{P(Result)} = \frac{P(Result\,|\,Cause)P(Cause)}{P(Result)}. \qquad (8.1)$$

This is just a mathematical theorem. Taking pattern recognition as an example, a concrete example is shown on how this theorem is used to estimate the cause.

The author would like to explain the problem of numbers pattern recognition by the Bayesian inference mentioned in the introduction as an example. Basically, it recognizes numbers from 0 to 9, but in principle, you should consider the problem of recognizing one number, for example, "2". It is to show various handwritten patterns of "2" and make them remember the characteristics of the pattern of "2".

Consider the following learning by considering a grid consisting of a total of $I \times J$ cells in rows I and columns J. It is assumed that each of the N handwritten patterns of "2" is drawn in blue color with a brush or pencil on a sheet of white paper. Set the above grid in the area where "2" is drawn. Let us represent the state of the cell in row i and column j as a binary variable that takes a binary value of 0 or 1. Assign 1 to blue cells and 0 to white cells.

By summing up the number of times 1 is given for each cell, the two-dimensional frequency distribution when it becomes 1 can be obtained. If the total number N of patterns is taken large enough, the discrete probability that cell is 1 for each cell can be defined by $P(C_{ij}=1\,|\,2),\ i=1,2,\cdots,I;\ j=1,2,\cdots,J$. $P(C_{ij}=0\,|\,2)=1-P(C_{ij}=1\,|\,2)$ is the probability that the cell is 0. The learning is to find this discrete probability distribution.

Next, consider the judgment based on the learning result from a viewpoint of the likelihood method. When some handwritten number n is shown, $C_{ij}=1$ when the cell is blue and $C_{ij}=0$ when it is white, as in the case of learning. The probability that it is "2" can be approximated:

$$P(C_{11},C_{12},\cdots,C_{IJ}\,|\,2)=\prod_{i=1}^{I}\prod_{j=1}^{J}P(C_{ij}\,|\,2). \tag{8.2}$$

This value will be greater when the numbers pattern is "2" than when it is not "2". Therefore, for example, when considering the recognition of handwritten digit patterns from 0 to 9, the above-mentioned learning first creates 10 types of discrete probability distributions corresponding to each digit. For judgment, the shown numerical pattern is shown in these 10 types of probability distributions. That is, if the calculation corresponding to equation (8.2) is performed, the calculated value corresponding to the indicated numbers will be the largest. From this, it will be possible to recognize what the handwritten numbers are.

The above can be expressed using mathematical formulas as follows. The conditional probability of the numbers pattern can be obtained approximately as

$$P(C_{11},C_{12},\cdots,C_{IJ}\,|\,n)=P(C_{11}\,|\,n)P(C_{12}\,|\,n)\cdots P(C_{IJ}\,|\,n). \tag{8.3}$$

If we apply Bayes' theorem, the reverse probability is given by

$$\begin{aligned}P(n\,|\,C_{11},C_{12},\cdots,C_{IJ})&=\frac{P(C_{11},C_{12},\cdots,C,n)}{P(C_{11},C_{12},\cdots,C_{IJ})}\\&=\frac{P(C_{11},C_{12},\cdots,C,n)}{\sum_{n=1}^{N}P(C_{11},C_{12},\cdots,C_{IJ},n)}=\frac{P(C_{11},C_{12},\cdots,C_{IJ}\,|\,n)P(n)}{\sum_{n=1}^{N}P(C_{11},C_{12},\cdots,C_{IJ}\,|\,n)P(n)}.\end{aligned} \tag{8.4}$$

The inference based on Bayes' theorem is called the Bayesian Inference. $P(n)$, $P(C_{11},C_{12},\cdots,C_{IJ}\,|\,n)$, and $P(n\,|\,C_{11},C_{12},\cdots,C_{IJ})$ are called the prior probability, the likelihood function, and the posterior probability, respectively.

When using Bayesian inference, the problem is always how to obtain prior probabilities. If you have some actual data, you can use it. If there is no such thing, you could be allowed generally to assume as follows:

$$P(0) = P(1) = \cdots = P(9) \tag{8.5}$$

If we assume equation (8.5), we have

$$P(n \mid C_{11}, C_{12}, \cdots, C_{IJ}) = \frac{P(C_{11}, C_{12}, \cdots, C_{IJ} \mid n)}{\sum_{n=1}^{N} P(C_{11}, C_{12}, \cdots, C_{IJ} \mid n)} \sim P(C_{11}, C_{12}, \cdots, C_{IJ} \mid n), \tag{8.6}$$

since the denominator does not depend on n.

This is nothing but the likelihood method. In the case of handwritten digit numbers recognition, the assumption in equation (8.5) may be considered appropriate. It could be possible to think that something similar to equation (8.6) is being performed in the human brain.

The above discussion positively incorporates the probability that the cell value will be 0, but if there is a cell with a cell value of 1 and a probability of 1, the probability of the cell with the cell value of 0 is 0. Cells with a cell value that is always zero have a variance of 0, so we use the probabilities excluding such cells to avoid the divergence of the calculation.

8.2. Application to recognition of simple numbers patterns [8]

8.2.1. Basic ideas of numbers pattern recognition

Although the following theory can be applied to arbitrary patterns, for the sake of simplicity, we consider digit pattern recognition.

The MNIST image data set of handwritten numbers from "0" to "9" [1] is converted into 0, 1 data in a 28×28 grid pattern. Firstly, let us consider the recognition of ten numeric patterns 0 to 9 drawn as a 5×5 grid pattern as shown in Fig. 8.1.

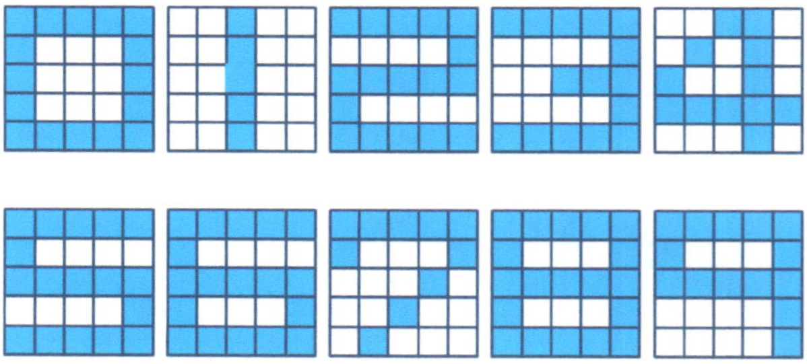

Fig. 8.1 Numbers patterns 0-9.

The prior probability of n = 0,1,...,9 numbers patterns is $P(n)$. The grid pattern shall consist of I rows and J columns. Row and column numbering is the same as for matrices. The top row is the first row and the bottom row is the Ith row. The leftmost column is the first column and the rightmost column is the Jth column.

The value of the *cell* (i, j), $i = 1,2,\ldots,I$, $j = 1,2,\ldots,J$ is C_{ij}, and when the cell is blue $C_{ij} = 1$, and when the cell is white, $Cell_{ij} = 0$ are set.

In the case of the numbers pattern n, it is assumed that the probability $P(C_{ij}|n)$ is known by learning. For example, for Fig. 8.1 it would be given as:

$$\left[P(C_{ij}=0|0)\right]=\begin{bmatrix}0&0&0&0&0\\0&1&1&1&0\\0&1&1&1&0\\0&1&1&1&0\\0&0&0&0&0\end{bmatrix},\quad \left[P(C_{ij}=1|0)\right]=\begin{bmatrix}1&1&1&1&1\\1&0&0&0&1\\1&0&0&0&1\\1&0&0&0&1\\1&1&1&1&1\end{bmatrix},$$

$$\left[P(C_{ij}=0|1)\right]=\begin{bmatrix}1&1&0&1&1\\1&1&0&1&1\\1&1&0&1&1\\1&1&0&1&1\\1&1&0&1&1\end{bmatrix},\quad \left[P(C_{ij}=1|1)\right]=\begin{bmatrix}0&0&1&0&0\\0&0&1&0&0\\0&0&1&0&0\\0&0&1&0&0\\0&0&1&0&0\end{bmatrix},\tag{8.7}$$

$$\cdots\cdots\cdots\cdots\cdots\cdots$$

$$\left[P(C_{ij}=0|9)\right]=\begin{bmatrix}0&0&0&0&0\\0&1&1&1&0\\0&0&0&0&0\\1&1&1&1&0\\0&0&0&0&0\end{bmatrix},\quad \left[P(C_{ij}=1|9)\right]=\begin{bmatrix}1&1&1&1&1\\1&0&0&0&1\\1&1&1&1&1\\0&0&0&0&1\\1&1&1&1&1\end{bmatrix},$$

In the case of recognition of handwritten numbers, many learning data are prepared and the probability distribution of blue and white of each cell is obtained for each numbers.

Then, the conditional probabilities $(C_{11},\ C_{12},\ldots,\ C_{IJ} \mid n)$ are approximated as follows:

$$P(C_{11},C_{12},\cdots,C_{IJ} \mid n) = P(C_{11} \mid n)P(C_{12} \mid n)\cdots P(C_{IJ} \mid n)\,. \tag{8.8}$$

Since the joint probability is given as

$$P(C_{11},C_{12},\cdots,C_{IJ},n) = P(C_{11},C_{12},\cdots,C_{IJ} \mid n)P(n)\,, \tag{8.9}$$

the inverse probability is given by

$$P(n \mid C_{11},C_{12},\cdots,C_{IJ}) = \frac{P(C_{11},C_{12},\cdots,C_{IJ},n)}{\sum_{n=1}^{N} P(C_{11},C_{12},\cdots,C_{IJ},n)}\,. \tag{8.10}$$

In the simple case as shown in Fig. 8.1, $P(C_{11},\ C_{12},\ldots,\ C_{IJ}|n)$ is calculated and normalized easily for n = 0-9. Then, it becomes the probability of each numbers. For example, if we input (11111 10000 11111 00001 11111) as shown in Fig. 8.2:

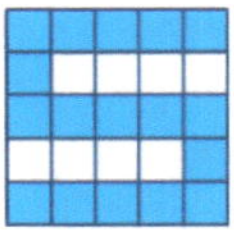

Fig. 8.2 Numbers pattern correctly recognized as 5.

the probability of n = 5 is 1 and the others are 0. For (01111 10000 11111 00001 11111) and (11111 11000 11111 00001 11111) as dhown in Fig. 8.3:

Fig. 8.3 Numbers patterns that are not correctly recognized as 5.

the probability of all n is 0.

It can be said that Bayesian learning recognizes as correctly as taught, to the extent that it can be called stupid. When generalization ability is required as in handwritten numeral recognition, it is necessary to learn a large amount of handwritten numeral data.

8.2.2. Simple figure pattern recognition (Checking of generalization ability)

When four horizontal bar patterns as shown in Fig. 8.4:

Pattern 1: Horizontal bar of length 5

Pattern 2: Length 4 horizontal bar to the left

Pattern 3: Length 4 horizontal bar to the right

Pattern 4: Horizontal bar of length 3

are recognized by Bayesian learning, the generalization ability of recognizing a given pattern accurately and recognizing a close pattern as one of given patterns is confirmed.

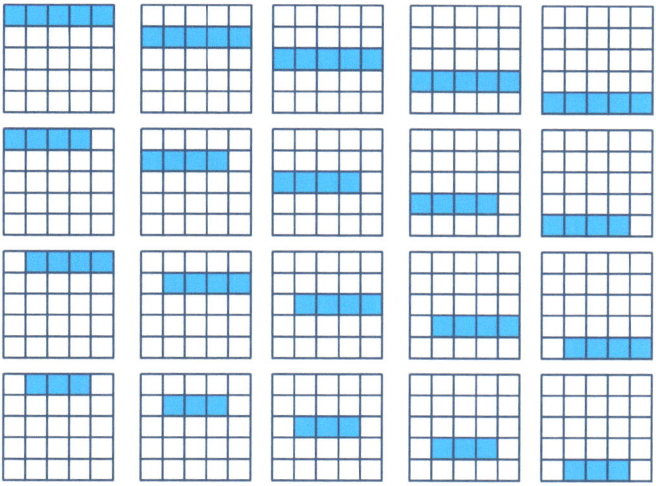

Fig. 8.4 Four horizontal bar patterns.

As a result of learning, the probability distribution that the value of the cell is 1 (Blue) is shown in Table 8.1 below. The probability distribution that the cell is 0 (White) is 1 minus the probability distribution that the cell is 1 (Blue).

Table 8.1 Probability distribution of each pattern.

(a) The probability distribution of cell value 1 for pattern 1:
$P(C_{11}, C_{12}, \ldots, C_{IJ} \mid \text{pat} = 1)$

	j=1	j=2	j=3	j=4	j=5
i=1	0.2	0.2	0.2	0.2	0.2
i=2	0.2	0.2	0.2	0.2	0.2
i=3	0.2	0.2	0.2	0.2	0.2
i=4	0.2	0.2	0.2	0.2	0.2
i=5	0.2	0.2	0.2	0.2	0.2

(b) The probability distribution of cell value 1 for pattern 2:
$P(C_{11}, C_{12}, \ldots, C_{IJ} \mid \text{pat} = 2)$

	j=1	j=2	j=3	j=4	j=5
i=1	0.2	0.2	0.2	0.2	0
i=2	0.2	0.2	0.2	0.2	0
i=3	0.2	0.2	0.2	0.2	0
i=4	0.2	0.2	0.2	0.2	0

i=5	0.2	0.2	0.2	0.2	0

(c) The probability distribution of cell value 1 for pattern 3:

$$P(C_{11}, C_{12}, \ldots, C_{IJ} \mid \text{pat} = 3)$$

	j=1	j=2	j=3	j=4	j=5
i=1	0	0.2	0.2	0.2	0.2
i=2	0	0.2	0.2	0.2	0.2
i=3	0	0.2	0.2	0.2	0.2
i=4	0	0.2	0.2	0.2	0.2
i=5	0	0.2	0.2	0.2	0.2

(d) The probability distribution of cell value 1 for pattern 1:

$$P(C_{11}, C_{12}, \ldots, C_{IJ} \mid \text{pat} = 4)$$

	j=1	j=2	j=3	j=4	j=5
i=1	0	0.2	0.2	0.2	0
i=2	0	0.2	0.2	0.2	0
i=3	0	0.2	0.2	0.2	0
i=4	0	0.2	0.2	0.2	0
i=5	0	0.2	0.2	0.2	0

It always recognizes teacher data correctly. For example, if the teacher data is input, it outputs correctly as the teacher data as shown in Table 8.2:

Table 8.2 Judgment of horizontal bar patterns which are teacher data.

N	Input Pattern	Recognition	No	Input Pattern	Recognition
1		Pattern 1	5	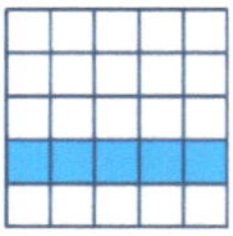	Pattern 1
2	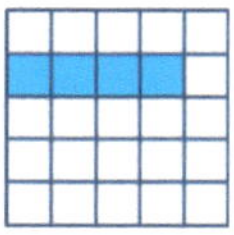	Pattern 2	6	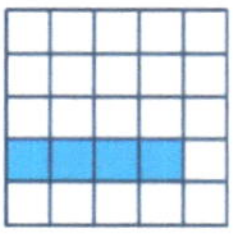	Pattern 2

3		Pattern 3	7	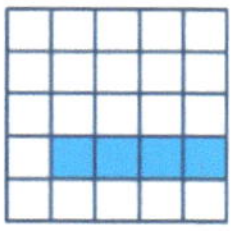	Pattern 3
4	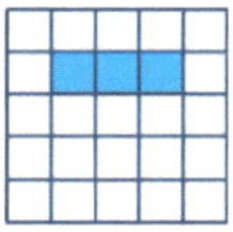	Pattern 4	8	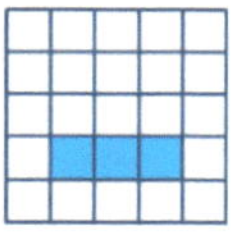	Pattern 4

If a horizontal bar pattern that is not teacher data is judged, it will be output as shown in, for example, Table 8.3, so the occurrence of generalization ability might be understood:

Table 8.3 Judgment of horizontal bar patterns that are not teacher data.

N	Input Pattern	Recognitio	No	Input Pattern	Recognitio
1		Pattern 1	6		Pattern 1
2	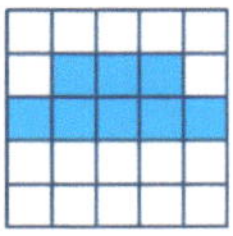	Pattern 1	7	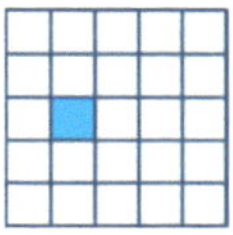	Pattern 4
3	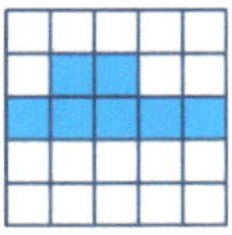	Pattern 1	8		Pattern 4
4	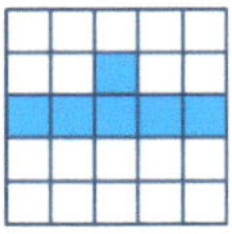	Pattern 1	9		Pattern 2
5		Pattern 1	10		Pattern 3

8.3. Application to recognition of MNIST hand-written numbers patterns [1,2,8,9]

8.3.1 MNIST hand-written numbers data

MNIST is an abbreviation of Modified National Institute of Standards and Technology, and the size of NIST data is arranged so that it is easy to use, and it is widely used for learning and evaluation of machine learning. The learning dataset was obtained from Census Bureau employees and the evaluation dataset was obtained from high school students. The original data is a black-and-white image, but it is a grayscale image with 0 to 255 gradations and is normalized so that it fits in the range of 28 × 28 pixels, so it is easy to use.

The MNIST database consists of 60,000 training images and 10,000 test images. Wikipedia [1] has a table of various classifiers and performances for your reference. This table is interesting because it shows low error rates due to various neural networks. By the way, the error rate of the deep neural network (DNN: 2-layer with 784-800-10 neurons in the input layer-intermediate layer-output layer) has a high accuracy of 1.6% without pretreatment. An extended dataset similar to MNIST called EMNIST was also released in 2017. This dataset contains handwritten numbers and letters, including 240,000 training images and 40,000 evaluation images.

MNIST data can be downloaded from the official MNIST page, but it is not easy. Reference [2] has a description of MNIST in CSV and PNG. For example, you can download CSV files for learning and testing by clicking mnist_train.csv (104 MB) or mnist_test.csv (17 MB).

When you open this file, each line is one handwritten numbers data, the cell in the first column is the label of each number, and the following cells 28 × 28 = 784 are filled with numbers from 0 to 255. Table 8.4 shows an example of the test data converted into a grid image.

Table 8.4 Image of test data

0	0	0	0	0	0	0	0	0	0	0	0	0	0	0	0	0	0	0	0	0	0	0	0	0	0	0	0
0	0	0	0	0	0	0	0	0	0	0	0	0	0	0	0	0	0	0	0	0	0	0	0	0	0	0	0
0	0	0	0	0	0	0	0	0	0	0	0	0	0	0	0	0	0	0	0	0	0	0	0	0	0	0	0
0	0	0	0	0	0	0	0	0	0	0	0	0	0	0	0	0	0	0	0	0	0	0	0	0	0	0	0
0	0	0	0	0	0	0	0	0	0	0	0	0	0	0	0	0	0	0	0	0	0	0	0	0	0	0	0
0	0	0	0	0	0	0	0	0	0	0	0	0	0	0	0	0	0	0	0	0	0	0	0	0	0	0	0
0	0	0	0	0	0	0	0	0	2	38	122	246	255	255	245	121	8	0	0	0	0	0	0	0	0	0	0
0	0	0	0	0	0	0	0	57	176	253	253	253	253	253	253	253	195	3	0	0	0	0	0	0	0	0	0
0	0	0	0	0	0	0	1	191	253	253	253	253	250	248	253	253	253	105	0	0	0	0	0	0	0	0	0
0	0	0	0	0	0	1	128	253	253	246	210	111	37	0	155	253	253	187	14	0	0	0	0	0	0	0	0
0	0	0	0	0	0	6	253	253	250	139	0	0	0	0	15	226	253	253	116	0	0	0	0	0	0	0	0
0	0	0	0	0	0	6	253	253	141	0	0	0	0	0	0	76	253	253	219	25	0	0	0	0	0	0	0
0	0	0	0	0	0	73	253	238	22	0	0	0	0	0	14	221	253	253	253	159	0	0	0	0	0	0	0
0	0	0	0	0	0	114	253	235	21	0	0	0	0	32	103	253	253	253	253	239	4	0	0	0	0	0	0
0	0	0	0	0	0	61	253	253	126	0	0	2	40	171	253	253	253	253	253	159	1	0	0	0	0	0	0
0	0	0	0	0	0	8	252	253	211	96	55	159	253	253	253	252	253	252	240	73	0	0	0	0	0	0	0
0	0	0	0	0	0	2	133	253	253	253	253	253	253	253	253	214	253	253	188	0	0	0	0	0	0	0	0
0	0	0	0	0	0	0	3	97	174	246	251	252	232	192	22	65	253	253	87	0	0	0	0	0	0	0	0
0	0	0	0	0	0	0	0	0	0	54	88	81	0	0	0	21	222	235	28	0	0	0	0	0	0	0	0
0	0	0	0	0	0	0	0	0	0	0	0	0	0	0	0	23	224	226	0	0	0	0	0	0	0	0	0
0	0	0	0	0	0	0	0	0	0	0	0	0	0	0	0	72	253	123	0	0	0	0	0	0	0	0	0
0	0	0	0	0	0	0	0	0	0	0	0	0	0	0	0	158	253	192	0	0	0	0	0	0	0	0	0
0	0	0	0	0	0	0	0	0	0	0	0	0	0	0	0	168	253	207	33	0	0	0	0	0	0	0	0
0	0	0	0	0	0	0	0	0	0	0	0	0	0	0	0	168	252	253	81	0	0	0	0	0	0	0	0
0	0	0	0	0	0	0	0	0	0	0	0	0	0	0	0	168	253	253	81	0	0	0	0	0	0	0	0
0	0	0	0	0	0	0	0	0	0	0	0	0	0	0	0	97	233	242	46	0	0	0	0	0	0	0	0
0	0	0	0	0	0	0	0	0	0	0	0	0	0	0	0	0	0	0	0	0	0	0	0	0	0	0	0
0	0	0	0	0	0	0	0	0	0	0	0	0	0	0	0	0	0	0	0	0	0	0	0	0	0	0	0

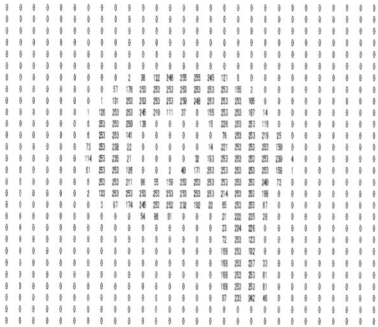

The above-mentioned Wikipedia [1] has a sample of handwritten numbers (16 images for each of 0 to 9), but some are unreadable by humans and some are interpreted differently by humans. We mentioned the high accuracy of deep neural networks, but We are worried that such high accuracy may have caused problems. I think that things that are difficult for humans to judge should be impossible to judge, rather than being mechanically forcibly judged.

8.3.2 MNIST handwritten numbers data learning and testing

In the following, learning and testing (judgment) of numbers pattern data from 0 to 9 by Bayesian inference using MNIST data will be described. Let c_{ij} be the value obtained by normalizing the gray scale of 255 gradations of cell (i, j) in a 28 × 28 grid with 255, and the conditional probability be $P(C_{ij} \mid n)$ when the number labels are from 0 to 10. We assume a normal distribution with mean $\mu_{n\,ij}$ and variance $v_{n\,ij}$. Namely,

$$P(C_{ij} \mid n) = \frac{1}{\sqrt{2\pi v_{n\,ij}}} \exp\left(-\frac{(C_{ij} - \mu_{n\,ij})^2}{2v_{n\,ij}} \right). \tag{8.11}$$

The mean $\mu_{n\,ij}$ and variance $v_{n\,ij}$ are obtained from the MNIST training data:

$$\mu_{n\,ij} = \frac{1}{M} \sum_{m=1}^{M} (C_{ij}{}^{m})_n, \quad v_{n\,ij} = \frac{1}{M} \sum_{m=1}^{M} \left[(C_{ij}{}^{m})_n - \mu_{n\,ij} \right]^2, \tag{8.12}$$

where $(C_{ij}{}^{m})_n$ means the m-th measured value of C_{ij} when the label is n. The equivalent of neural network learning is nothing but finding the mean $\mu_{n\,ij}$ and variance $v_{n\,ij}$, which are the parameters of the probability distribution, from the training data using equation (8.12). Therefore, the meaning of learning is clear, and it has nothing to do with overfitting in the case of neural networks.

Next, as an inference at the time of the test, if the label is n, the probability (likelihood function) $P(C_{11}, C_{12}, \cdots, C_{27\,28}, C_{28\,28} \mid n)$ that the test data of the value $(C_{11}, C_{12}, \cdots, C_{27\,28}, C_{28\,28})$ appears is approximated:

$$\begin{aligned} &P(C_{11}, C_{12}, \cdots, C_{27\,28}, C_{28\,28} \mid n) \\ &= P(C_{11} \mid n) P(C_{12} \mid n) \cdots P(C_{28\,27} \mid n) P(C_{28\,28} \mid n) = \prod_{i=1}^{28} \prod_{j=1}^{28} P(C_{il} \mid n) \end{aligned} . \tag{8.13}$$

If the prior probability of the label n is $P(n)$, the joint probability $P(C_{11}, C_{12}, \cdots, C_{27\,28}, C_{28\,28})$ is

$$P(C_{11}, C_{12}, \cdots, C_{27\,28}, C_{28\,28}, n) = P(C_{11}, C_{12}, \cdots, C_{27\,28}, C_{28\,28} \mid n) P(n), \tag{8.14}$$

the posteriori probability is given by

$$P(n \mid C_{11}, C_{12}, \cdots, C_{27\,28}, C_{28\,28}) = \frac{P(n, C_{11}, C_{12}, \cdots, C_{27\,28}, C_{28\,28})}{\sum_{n=1}^{10} P(n, C_{11}, C_{12}, \cdots, C_{27\,28}, C_{28\,28})}$$
$$= \frac{P(C_{11}, C_{12}, \cdots, C_{27\,28}, C_{28\,28} \mid n) P(n)}{\sum_{n'=1}^{10} P(C_{11}, C_{12}, \cdots, C_{27\,28}, C_{28\,28} \mid n') P(n')}. \tag{8.15}$$

Furthermore, since we could assume

$$P(0) = P(1) = \cdots = P(9), \tag{8.16}$$

the condition of maximum a posteriori is equal to the condition of maximum likelihood function:

$$\text{Estimation of } n = \max_n P(C_{11}, C_{12}, \cdots, C_{27\,28}, C_{28\,28} \mid n). \tag{8.17}$$

8.3.3 Numerical calculation result

In numerical calculation, in order to avoid underflow, the calculation of Eq. (8.13) is performed by taking the natural logarithm of both sides. This is because the order of the numerical magnitude does not change even if the logarithm is taken. That is

$$\ln P(C_{11}, C_{12}, \cdots, C_{27\,28}, C_{28\,28} \mid n)$$
$$= \ln P(C_{11} \mid n) + \ln P(C_{12} \mid n) + \cdots + \ln P(C_{28\,27} \mid n) P(C_{28\,28} \mid n) \tag{8.18}$$
$$= \sum_{i=1}^{28} \sum_{j=1}^{28} \ln P(C_{il} \mid n).$$

Hence, we obtain

$$\text{Estimation of } n = \max_n \ln P(C_{11}, C_{12}, \cdots, C_{27\,28}, C_{28\,28} \mid n). \tag{8.19}$$

Figure 8.5 shows a part of the results of finding the mean $\mu_{n\,ij}$ and variance $v_{n\,ij}$, which are the parameters of the probability distribution obtained by training using all 60,000 training data. The calculation to find these takes only a few minutes on a personal computer.

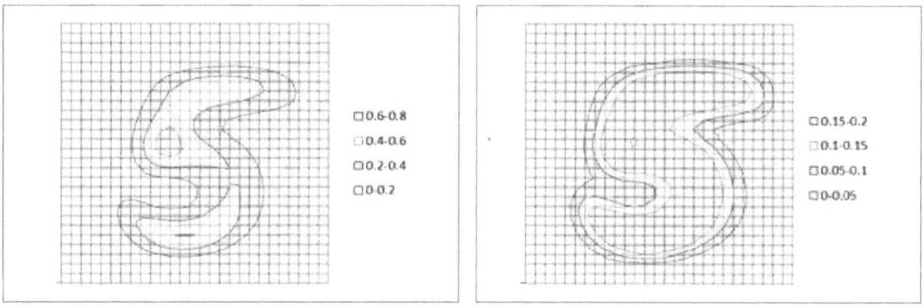

Fig. 8.5 Distributions of mean $\mu_{n\,ij}$ and variance $v_{n\,ij}$ of label 5

There is one difficult point about this problem. It is the handling of the white background. Since $C_{ij}=0$ in the white background, some cells will always be zero in all training data. In such a cell, the mean $\mu_{n\,ij}=0$ and the variance $v_{n\,ij}=0$ according to equation (8.12). As shown in Fig. 8.5, the range with $v_{n\,ij}=0$ is quite wide. This means that when evaluating the likelihood function with equation (8.18), the likelihood function diverges due to equations (8.15) and (8.11). The measures to avoid this will be described in detail in §8.3.4, but one idea is to limit the range of taking the product in equation (8.13) to the cell with $v_{n\,ij}\neq 0$.

Table 2 shows the results of making the first 100 judgments out of 10,000 test data according to the above idea. Unfortunately, the correct answer rate is about 80%, which is not very high. The reason why the correct answer rate is not high is that the total number of cells is 28 × 28 = 784, and the degree of freedom of judgment is only 2 × 784, which is the mean and variance of each cell. In the neural network described in §8.3.1, there are 784 × 800 = 627,200 weights between the input layer and the intermediate layer. If the one-hot part of the top layer is included, 800 x 10 = 8,000 weights are added to this.

Table 8.5 Results of the test (The asterisk refers to the wrong result.)

test_no.	test_lbl	n_max	test_no.	test_lbl	n_max	test_no.	test_lbl	n_max	test_no.	test_lbl	n_max	test_no.	test_lbl	n_max
1	7	7	21	9	9	41	1	1	61	7	7	81	7	7
2	2	2	22	6	6	42	7	7	*62	8	2	82	6	6
3	1	1	23	6	6	43	4	4	*63	9	7	83	2	2
4	0	0	24	5	5	44	2	2	*64	3	2	84	7	7
5	4	4	25	4	4	45	3	3	65	7	7	*85	8	4
6	1	1	26	0	0	*46	5	3	*66	4	1	86	4	4
7	4	4	27	7	7	*47	1	3	*67	6	2	87	7	7
8	9	9	28	4	4	*48	2	6	68	4	4	88	3	3
*9	5	4	29	0	0	49	4	4	69	3	3	89	6	6
*10	9	7	30	1	1	50	4	4	70	0	0	90	1	1
11	0	0	31	3	3	51	6	6	71	7	7	91	3	3
12	6	6	32	1	1	52	3	3	72	0	0	92	6	6
13	9	9	33	3	3	53	5	5	73	2	2	*93	9	1
14	0	0	*34	4	0	*54	5	3	*74	9	7	94	3	3
15	1	1	35	7	7	55	6	6	75	1	1	95	1	1
*16	5	3	36	2	2	56	0	0	76	7	7	96	4	4
17	9	9	37	7	7	57	4	4	77	3	3	97	1	1
18	7	7	38	1	1	58	1	1	78	2	2	*98	7	2
*19	3	2	39	2	2	59	9	9	79	9	9	99	6	6
20	4	4	40	1	1	*60	5	4	80	7	7	100	9	9

8.3.4 Measures for white background treatment and improvement of judgment accuracy

(1) Treatment of white background

In §8.3.3, the importance of treating white backgrounds was described. For cells with zero variance $v_{n\,ij}=0$ of training data, instead of equation (8.11), one could think as follows, for example:

$$P(C_{ij}\,|\,n)=\begin{cases}\ln 2, & \text{if } C_{ij}=0\\ \ln 0.01, & \text{otherwise}\end{cases}. \tag{8.20}$$

In this case, the correct answer rate was 79.69% for 10,000 test data.

You can also add 0.01 to all the variances of the training data so that $v_{n\,ij}=0$ does not occur. The test data is left as given. At this time, the correct answer rate was 81.43% for 10,000 test data.

It is conceivable to make the training data virtual so that it does not become $v_{n\,ij}=0$. For example, for a white background:

$$C_{n\,ij}=0.35\times(\text{drand}()+1.0)\times 0.5\times 255, \tag{8.21}$$

where, $\text{drand}()$ is a function that generates a uniform random number of (-1,1). In this case, the correct answer rate was 84.96% for 10,000 test data.

Instead of equation (8.21), we could assume

$$C_{n\,ij}=0.3\times|(\text{rand\_normal}(0.5, 0.5+0.125\times\text{drand}())|\times 255, \tag{8.22}$$

where, $\text{rand\_normal}(\mathit{mean},\mathit{sigma})$ is a function that generates a normal random number with mean *mean* and standard deviation *sigma*. It takes an absolute value so that is not negative. In this case, the correct answer rate was 85.17% for 10,000 test data.

The programing code is given in Appendix8A.

(2) Increased degree of freedom in probability distribution

To increase the degrees of freedom of the probability distribution, the approximations given by equations (8.11) and (8.13) should be replaced with the original multivariate normal distribution:

$$P(C_{11},C_{12},\cdots,C_{27\,28},C_{28\,28}\,|\,n)=\frac{\exp\left(-\frac{1}{2}(\mathbf{C}-\boldsymbol{\mu})^T\boldsymbol{\Sigma}^{-1}(\mathbf{C}-\boldsymbol{\mu})\right)}{\sqrt{(2\pi)^{784}\,|\boldsymbol{\Sigma}|}}, \tag{8.23}$$

where, $\mathbf{C}=(C_{11},C_{12},\cdots,C_{27\,28},C_{28\,28})^T$ and $\boldsymbol{\mu}=(\mu_{11},\mu_{12},\cdots,\mu_{27\,28},\mu_{28\,28})^T$ are 28 × 28 = 784-dimensional column vectors, and $\boldsymbol{\Sigma}$ is a 784 × 784 covariance matrix. $|\boldsymbol{\Sigma}|=\det\boldsymbol{\Sigma}$ is the determinant of $\boldsymbol{\Sigma}$. The mean $\mu_{i\,j}$ of cell (i, j) can be calculated from equation (8.12), but the covariance $v_{n\,ijkl}$ of cell (i, j) and cell (k,l) is a multidimensional extension of the one-dimensional variance $v_{n\,ij}$:

$$v_{n\,ijkl} = \frac{1}{M}\sum_{m=1}^{M}\left[(C_{ij}{}^{m})_n - \mu_{n\,ij}\right]\left[(C_{kl}{}^{m})_n - \mu_{n\,kl}\right]. \tag{8.24}$$

Diagonal approximation is assumed for the covariance determinant $|\mathbf{\Sigma}|$:

$$|\mathbf{\Sigma}| \Box \prod_{i=1}^{28}\prod_{j=1}^{28} v_{n\,ij} \tag{8.25}$$

To avoid underflow, consider the natural logarithm of both sides of equation (18):

$$\begin{aligned} &\ln P(C_{11}, C_{12}, \cdots, C_{27\,28}, C_{28\,28} \,|\, n) \\ &= -\frac{1}{2}\left[(\mathbf{C}-\boldsymbol{\mu})^T \mathbf{\Sigma}^{-1}(\mathbf{C}-\boldsymbol{\mu}) + 784 \times \ln(2\pi) + \ln|\mathbf{\Sigma}|\right]. \end{aligned} \tag{8.26}$$

It is possible to make the training data virtual so that it does not become $v_{n\,ij} = 0$. For example, for a white background:

$$C_{n\,ij} = 0.3 \times |(\text{rand\_normal}(0.5 + 0.5 \times \text{drand}(), 0.5 + 0.125 \times \text{drand}())| \times 255, \tag{8.22}$$

where, $\text{rand\_normal}(mean, sigma)$ is a function that generates a normal random number with mean *mean* and standard deviation *sigma*. It takes an absolute value so that is not negative. In this case, the correct answer rate was 95.65% for 10,000 test data.

Prior probabilitie $P(n)$ could also affect accuracy. We also need to know this effect. In the case of this problem, a large amount of training data of 60,000 is used, so prior probabilities can be estimated through the training data. The results are shown in Table 8.6, but since they are almost the same with $P(n) \approx 0.1$, the effect is considered to be small. The accuracy rate did not change much even when the calculation was performed by introducing the prior probabilities estimated from the training data.

Table 8.6 Prior probability estimated using train data.

n	Frequency	Prior Prob.
0	5923	0.098717
1	6742	0.112367
2	5958	0.0993
3	6131	0.102183
4	5842	0.097367
5	5421	0.09035
6	5918	0.098633

7	6265	0.104417
8	5851	0.097517
9	5949	0.09915

How much the correct answer rate can be raised is an issue for the future.

8.5. Conclusions

With the advent of deep learning, breakthroughs have occurred in AI technologies such as image recognition, voice recognition, and machine translation that humans can easily do but machines cannot. However, deep learning is not the only AI technology. There are problems such as taking a long time to learn and abnormal recognition due to over-fitting, but the biggest problem is anxiety that comes from the fact that the reasoning is a black box and the basis of recognition is not known sufficiently.

Since Bayesian learning is based on Bayesian inference in statistics, it is based on a theory completely different from deep learning. Since the learning method is completely different, it has the potential to free us from the problems of deep learning described above.

Neural networks are regarded as mathematical models of information processing in the cerebrum. As mentioned in section 2, the Bayesian inference is also considered to suggest information processing different from neural networks in the cerebrum. The Bayesian inference is interesting in that sense.

Learning in the Bayesian inference is quite different from learning in neural networks. Learning in the Bayesian inference is to find the probability of a chosen random variable from a large number of data. In addition, the judgment in the Bayesian inference is to judge what kind of data is required based on the probability distribution obtained by learning.

In this study, we applied the Bayesian inference to the number pattern recognition. This problem has been solved by the neural network, but we have dealt with what happens from the perspective of the Bayesian inference. I hope that new developments will be born by looking at it from a new perspective.

REFERENCES 8

[1] Wikipedia, MNIST database, https://en.wikipedia.org/wiki/MNIST_database.

[2] Joseph Redmon, MNIST in CSV and PNG, https://github.com/pjreddie/mnist-csv-png (2018).

[3] Krizhevsky, A., Sutskever, I. & Hinton, G. ImageNet classification with deep convolutional neural networks. In Proc. Advances in Neural Information Processing Systems 25 1090–1098 (2012).

[4] Yann LeCun1, Yoshua Bengio & Geoffrey Hinton, Deep learning, NATURE | VOL 521 | 28 MAY (2015).

[5] Ian Goodfellow, Yoshua Bengio and Aaron Courville, Deep Learning, The MIT Press 2017.

[6] A. Wald, Statistical Decision Functions, John Wiley & Sons, Inc. 1956.

[7] M. Antonia Amaral Turkman, Carlos Daniel Paulino, Peter Muller, Computational Bayesian Statistics, Cambridge Univ. Press, 2019.

[8] Hiroshi Isshiki, Pattern Recognition by Bayesian Inference, The 63rd Joint Meeting of Automatic Control (2020). https://www.jstage.jst.go.jp/article/jacc/63/0/63_329/_pdf

[9] Hiroshi Isshiki, Pattern Recognition by Bayesian Inference Using MNIST Data, 2002 JSME Annual Meeting, (2022). Publication pending, in Japanese.

Appendix 8A C-language code for MNIST number pattern recognition

"Microsoft C/C++ Compiler Version 17.00.50727.1 for x86" and "Microsoft Linker Version 11.00.50727" were used for compile and link (in command window; cl source_file_name.c).

(1) Progrming code: BayesMNIST_testAllNorm.c

```
//--------------------------------------------------------------------------------//
//                                                                                //
// File Name: BayesMNIST.c                               2022.03.24-2022.03.27 //
//            BayesMNISTX.c                              2022.03.27-2022.03.27 //
//            BayesMNIST_train.c                         2022.03.27-2022.03.29 //
//            BayesMNIST_test.c                          2022.03.29-2022.03.31 //
//            BayesMNIST_testAll.c                       2022.03.31-2022.04.07 //
//            BayesMNIST_testAllNorm.c                   2022.04.10-2022.04.12 //
//                                                                                //
//    MNIST bumbrt pattern                                                        //
//    Treatment of white gackground for training data :                           //
//                Replaced by tmp = alp*rand_normal(myu, bet+gam*drand())*255.0   //
//    Test data were not changed.                                                 //
//                                                                                //
//--------------------------------------------------------------------------------//

#include <stdio.h>
```

```
#include <stdlib.h>
#include <string.h>
#include <math.h>

#define PI      3.14159265                          // Pi

// ----- functions ------------------------------------------------------ //

void main();
void pushKey();

double normal();                                    // normal distribution

double Uniform( void );                             // uniform random number
double rand_normal(double,double);                  // normal random number

double drand();                                     // uniform random number in (-1,1)

// ----- variables ------------------------------------------------------ //

char title_memo[5000];

long N;                          // number of trraining data
int M;                           // number of test data

double alp;                      // tmp = alp*rand_normal(myu,bet+gam*drand())*255.0
double myu;                      // tmp = alp*rand_normal(myu,bet+gam*drand())*255.0
double bet;                      // tmp = alp*rand_normal(myu,bet+gam*drand())*255.0
double gam;                      // tmp = alp*rand_normal(myu,bet+gam*drand())*255.0

long n;                          // number
long m;                          // number
long m1;                         // number
int num_cor;                     // number of correct judgement

int Pat[29][29];                 // CSVdata
```

```
int Lbl[10];                    // number of label 0,1,2,...,9

int lbl;                        // label

int lbl1[10001];                // label for the m1-th test data
int m1Max[10001];               // estimated label for the m1-th test data
double proMax[10001];           // max probability for the m1-th test data

double avg[10][29][29];         // average
double vtmp[10][29][29];        // temporary
double var[10][29][29];         // variance
long nn[10][29][29];            // number of each label;

double cor[10];                 // correlation between sample and teacher
double pro[10];                 // p(sample | teacher)

double AMAT[1001][2001];        // matrix

FILE *fp_inp;                   // file pointer of input file
FILE *fp_out;                   // file pointer of output file

char InputDataFile[80];         // input file name
char OutputDataFile[80];        // output file name

char buf[5000];

// -------------------------------------------------------------- //

void main()
{
    int i, j;
    double tmp;

    //// open input file for parameters
```

```
sprintf(InputDataFile, "BayesMNIST_testAllNorm_inp.dat");

if ((fp_inp = fopen(InputDataFile, "r")) == NULL) {
    printf("Failed in Reading Input Data File! ... %s¥n", InputDataFile);
    exit(1);
}

//// open output file
sprintf(OutputDataFile, "BayesMNIST_testAllNorm_out.csv");  //////////

if ((fp_out = fopen(OutputDataFile, "w")) == NULL) {
    printf("Failed in Reading Output Data File! ... %s¥n", OutputDataFile);
    exit(1);
}

//// input from file
fscanf(fp_inp, "%s", title_memo);

fscanf(fp_inp, "%s %d", buf, &N);
fscanf(fp_inp, "%s %d", buf, &M);

fscanf(fp_inp, "%s %lf", buf, &alp);
fscanf(fp_inp, "%s %lf", buf, &myu);
fscanf(fp_inp, "%s %lf", buf, &bet);
fscanf(fp_inp, "%s %lf", buf, &gam);

fclose(fp_inp);

//// output to display
printf("memo: %s¥n", title_memo);

printf("N      = %d¥n", N);
printf("M      = %d¥n", M);
```

```
printf("alp   = %12.6f¥n", alp);
printf("myu   = %12.6f¥n", myu);
printf("bet   = %12.6f¥n", bet);
printf("gam   = %12.6f¥n", gam);
printf("¥n");


//// ouput to file
fprintf(fp_out, "memo: %s¥n", title_memo);
fprintf(fp_out, "¥n");

fprintf(fp_out, "N =, %d¥n", N);
fprintf(fp_out, "M =, %d¥n", M);

fprintf(fp_out, "alp =, %12.6f¥n", alp);
fprintf(fp_out, "myu =, %12.6f¥n", myu);
fprintf(fp_out, "bet =, %12.6f¥n", bet);
fprintf(fp_out, "gam =, %12.6f¥n", gam);
fprintf(fp_out, "¥n");

pushKey();

//////////////////////////////
//                          //
//          train           //
//                          //
//////////////////////////////

//// open input file of training data
sprintf(InputDataFile, "mnist_train.csv");

if ((fp_inp = fopen(InputDataFile, "rt")) == NULL) {
    printf("Failed in Reading Input Data File! ... %s¥n", InputDataFile);
    exit(1);
```

```
    }
    // calculation of mean and variance of each label and cell
    for (n = 1; n <= N; n++) {
        fscanf(fp_inp, "%d, ", &lbl);

        for (i = 1; i <= 28; i++)
            for (j = 1; j <= 28; j++) {
                fscanf(fp_inp, "%d, ", &Pat[i][j]);

                if (Pat[i][j] == 0) {
                    tmp = alp*rand_normal(myu,bet+gam*drand())*255.0;
                   if (tmp < 0)
                        tmp = -tmp;
                    Pat[i][j] = tmp;
                }
            }

        for (i = 1; i <= 28; i++)
            for (j = 1; j <= 28; j++) {
                nn[lbl][i][j] += 1;
                avg[lbl][i][j] += (Pat[i][j]+0.0)/255.0;
                vtmp[lbl][i][j] += (Pat[i][j]+0.0)/255.0*(Pat[i][j]+0.0)/255.0;
            }
    }

    fclose(fp_inp);

    for (lbl = 0; lbl <= 9; lbl++)
        for (i = 1; i <= 28; i++)
            for (j = 1; j <= 28; j++) {
                avg[lbl][i][j] /= nn[lbl][i][j]+0.0;
                var[lbl][i][j] = vtmp[lbl][i][j]/(nn[lbl][i][j]+0.0) -
avg[lbl][i][j]*avg[lbl][i][j];
            }
```

```
//// output 2 to file...mean of each label and cell
for (lbl = 0; lbl <= 9; lbl++) {
    fprintf(fp_out, "avg, lbl =, %d¥n", lbl);

    // upside down //
    for (i = 28; i >= 1; i--) {
        for (j = 1; j <= 28; j++)
            fprintf(fp_out, "%12.6f, ", avg[lbl][i][j]);
        fprintf(fp_out, "¥n");
    }
    fprintf(fp_out, "¥n");
}
fprintf(fp_out, "¥n");

//// output 2 to file...variance of each label and cell
for (lbl = 0; lbl <= 9; lbl++) {
    fprintf(fp_out, "var, lbl =, %d¥n", lbl);

    // upside down //
    for (i = 28; i >= 1; i--) {
        for (j = 1; j <= 28; j++)
            fprintf(fp_out, "%12.6f, ", var[lbl][i][j]);
        fprintf(fp_out, "¥n");
    }
    fprintf(fp_out, "¥n");
}
fprintf(fp_out, "¥n");


//////////////////////////////
//                          //
//          test            //
//                          //
//////////////////////////////
```

```
//// open input file of test data
sprintf(InputDataFile, "mnist_test.csv");

if ((fp_inp = fopen(InputDataFile, "rt")) == NULL) {
    printf("Failed in Reading Input Data File! ... %s¥n", InputDataFile);
    exit(1);
}

for (m1 = 1; m1 <= M; m1++) {

    fscanf(fp_inp, "%d, ", &lbl);
    for (i = 1; i <= 28; i++)
        for (j = 1; j <= 28; j++)
            fscanf(fp_inp, "%d, ", &Pat[i][j]);

    fprintf(fp_out, "m1 =, %d¥n", m1);
    fprintf(fp_out, "lbl =, %d¥n", lbl);

    // upside down //
    for (i = 28; i >= 1; i--) {
        for (j = 1; j <= 28; j++)
            fprintf(fp_out, "%d, ", Pat[i][j]);
        fprintf(fp_out, "¥n");
    }
    fprintf(fp_out, "¥n");



    // probability
    for (m = 0; m <= 9; m++) {
        pro[m] = 0.0;
        for (i = 1; i <= 28; i++)
            for (j = 1; j <= 28; j++)
                pro[m] += -0.5*log(2.0*PI*(var[m][i][j]+0.0))        //////////
```

```
                    -
((Pat[i][j]+0.0)/255.0-avg[m][i][j])*((Pat[i][j]+0.0)/255.0-avg[m][i][j])/2.0/(var[m][i][j
]+0.0);
        }

        printf(" m1 =,  %d¥n", m1);
        printf(" lbl =,  %d¥n", lbl);
        printf("m, pro[m]¥n");
        for (m = 0; m <= 9; m++)
            printf("%d, %e¥n", m, pro[m]);
        printf("¥n");

        fprintf(fp_out, " m1 =,  %d¥n", m1);
        fprintf(fp_out, " lbl =,  %d¥n", lbl);
        fprintf(fp_out, " m, pro[lbl, m]¥n");
        for (m = 0; m <= 9; m++)
            fprintf(fp_out, "%d, %e¥n", m, pro[m]);
        fprintf(fp_out, "¥n");

        lbl1[m1] = lbl;
        proMax[m1] = -1000000000.0;
        m1Max[m1] = 0;
        for (m = 0; m <= 9; m++)
            if (proMax[m1] < pro[m]) {
                m1Max[m1] = m;
                proMax[m1] = pro[m];
            }

    }

    num_cor = 0;
    fprintf(fp_out, "***** Result of Numbder Pattern Recognition *****¥n");
    fprintf(fp_out, "m1, lbl1, m1Max, proMax¥n");
    for (m1 = 1; m1 <= M; m1++) {
```

```
        fprintf(fp_out, "%ld, %d, %d, %12.6f,", m1, lbl1[m1], m1Max[m1], proMax[m1]);
        if (lbl1[m1] == m1Max[m1]) {
            fprintf(fp_out,"¥n");
            num_cor++;
        }
        else
            fprintf(fp_out,"*¥n");
    }
    fprintf(fp_out, "¥n");
    printf("cor pct =, %12.6f¥n", (num_cor+0.0)/(M+0.0)*100.0);
    fprintf(fp_out, "cor pct =, %12.6f¥n", (num_cor+0.0)/(M+0.0)*100.0);

    fclose(fp_out);

    pushKey();
}

// ---------------------------------------------------------------- //

void pushKey()
{
    printf("¥n        Push Return Key! ");
    getchar();
    getchar();
}

// ---------------------------------------------------------------- //

double normal(double x, double myu, double sgm)
{
    return 1.0/sqrt(2.0*PI*sgm*sgm)*exp(-(x-myu)*(x-myu)/(2.0*sgm*sgm));
}

// ---------------------------------------------------------------- //
```

```
double Uniform( void )
{
  static int x=10;
  int a=1103515245,b=12345,c=2147483647;
  x = (a*x + b)&c;

  return ((double)x+1.0) / ((double)c+2.0);
}

// --------------------------------------------------------------- //

double rand_normal( double myu, double sigma )
{
  double z=sqrt( -2.0*log(Uniform()) ) * sin( 2.0*PI*Uniform() );
  return myu + sigma*z;
 }

// --------------------------------------------------------------- //

double drand()
{
     return 2.0*(((double)rand())/((double)RAND_MAX)-0.5);
}

// --------------------------------------------------------------- //
```

(2) Input data: BayesMNIST_testAllNorm_inp.dat

```
2022.04.12

N:train          60000
M:test           10000

alp              0.3
```

myu	0.5
bet	0.5
gam	0.125

9. INFERENCE IN THE BRAIN: MINIMUM PRINCIPLE OF FREE ENERGY

Friston's variational free energy minimum principle [1] seems to be an important academic idea. However, it is mathematically advanced and Friston's explanation is redundant and difficult to understand accurately. There are excellent explanations by Bukley et al. [2] and Bogacz [3]. The former is characterized by dynamic aspects, and the latter is characterized by learning and bio-implementation. In this book, we explain this principle from a bit different viewpoint and show its application examples through numerical calculation. Especially for dynamic problems, Friston's explanation [4] is complicated and difficult to understand. We describes a much easier way of thinking [5]. The same calculation as Buckley [2] is actually performed, and the same result is obtained.

9.1 Helmholtz viewpoint of inference in brain

As is well known, various so-called minimum value principles are known for phenomena in physics and chemistry. For example, the principle of Least Action, which is famous in mechanics. Historically, it has been established under the name of the minimum principles, but strictly speaking, they should be called the stationary principles or the variational principles. Friston proposed that such a minimum principle also applies to biological phenomena such as information processing in the cerebrum. That is what is called Friston's variational free energy minimum principle [1]. The meaning of variational here will be described later.

Let us consider the fundamental reason why such a principle holds. It can be said that the greatest purpose of biological activity is to stay alive, that is, self-preservation. Most self-preservation is the maintenance of equilibrium. Maintaining equilibrium, according to Helmholtz, is thermodynamically maintaining the minimum free energy state. The state of the minimum free energy changes according to the change of the external environment, and adapts to the environment. Helmholtz is a physiologist and physicist representing Germany in the 19th century. Free energy is mechanical energy that can be extracted in the isothermal and isochoric state, and in the isothermal and isochoric state, chemical changes are led spontaneously in the direction of decreasing free energy.

The cerebrum, which is a part of living organisms, is responsible for information processing, which is an important element of biological activity. Various information processing is performed according to the change of the external environment, that is, the internal environment is changed, and the change is considered to be led to the direction in which a new equilibrium state is realized and follows the minimum free energy. In other words, it seems to occur in the direction of minimizing the surprise. This is considered to be Friston's claim. In other words, it seems to occur in the direction of minimizing the surprise. This is considered to be Friston's claim.

Brain reasoning is thought to occur as described above, but Helmholtz's important point is still important for brain reasoning itself. His viewpoint "what you see is unconscious reasoning" is very important. It sounds unbelievable at first, but it's a really important point. We tend to think that we are looking at the image itself reflected on the retina. However, what we think we are seeing is not the image of the retina itself, but the result of inference by the cerebrum based on the retinal image.

When we see letters in a foreign language that we have never learned or figures that we have never seen, our brains start to work hard and try to find meaning. The brain keeps inferring until it finds a convincing meaning. Everyone experiences it every day. It seems to be a major feature of our brain that we always seek some kind of interpretation for external stimuli and repeat inferences.

Moreover, although the landscape reflected on the retina is only two pieces of two-dimensional images, we regard the landscape as a three-dimensional image inferred from these two pieces of two-dimensional images. Another example is eye deceiving pictures. For example, what appears to be a symmetrical black vase is drawn on white paper [6]. At first, we think it was a vase, but when I pay attention to the white part, we can see the faces of the two persons facing each other.

Another example is the rubber hand illusion [7]. Place your hands on the table with your palms facing down. Open your hands wider than your shoulders and stand an opaque screen next to your left hand to hide your left hand. Place a rubber hand made of rubber on the opposite side of the opaque screen. Another person uses a brush to stimulate the fingers of the left and left rubber hands at the same time. After stimulating several times, the rubber hand becomes an illusion of a real left hand. At this stage, if you hit the rubber hand with a hammer, he will scream as if his real left hand was hit by the hammer

9.2 Variational free energy

9.2.1 Estimation of the length of the bar

In discussing brain problems, we make a clear distinction between the environment outside the body and the inside of the body. Solving a problem means finding the posterior probability of inferring the cause from the result. Bayesian inference is replaced by the minimum problem in the brain. Free energy is minimized. Mathematically, the variational free energy is minimized because the approximate solution of the posterior probability is obtained.

Figure 9.1 compares measuring the length of the bar with estimating the length of the bar in the brain. In both cases, measuring or estimating the length of the bar has been shown to be nothing more than finding posterior probabilities.

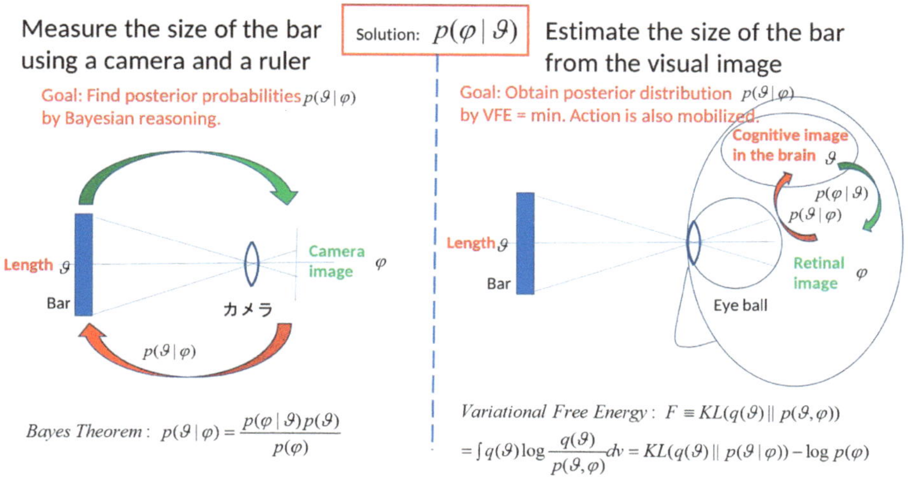

Fig. 9.1 The contrast between measuring the length of a rod with a camera and measuring it by eye.

Let's dig a little deeper into the problem of measuring the length of a bar. As a result of the measurement, suppose that the data shown in Fig. 9.2 is obtained. Let's estimate the length of the bar from the N measured values $(\varphi_1, \varphi_2, \cdots, \varphi_N)$. Consider the least squares method, maximum likelihood method, and Bayesian inference.

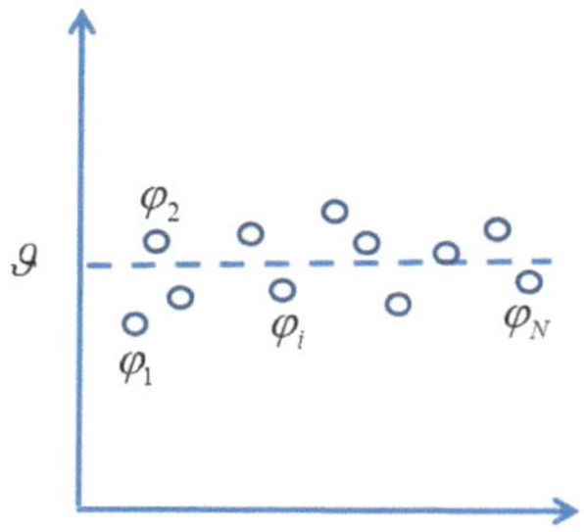

Fig. 9.2 Measured data of the bar

The observation equation in this case is given by:

$$\vartheta = \varphi_i + \varepsilon \,. \tag{9.1}$$

First, the least squares method is used, and the error is E:

$$E \equiv \frac{1}{2}\sum_{i=1}^{N}(\vartheta - \varphi_i)^2 \,. \tag{9.2}$$

Since we get from this:

$$\vartheta = \arg\min_{\vartheta} E \rightarrow \frac{\partial E}{\partial \vartheta} = \sum_{i=1}^{N} (\vartheta - \varphi_i) = 0, \tag{9.3}$$

ϑ is estimated as follows:

$$\vartheta = \frac{1}{N} \sum_{i=1}^{N} \varphi_i \,. \tag{9.4}$$

For stochastic inferences such as maximum likelihood method and Bayesian inference, consider the following conditional and prior probabilities as a normal distribution in preparation:

$$p(\boldsymbol{\varphi} \mid \vartheta) = \prod_{i=1}^{N} \frac{1}{\sqrt{2\pi\Sigma_{\varphi}}} \exp\left(-\frac{(\varphi_i - \vartheta)^2}{2\Sigma_{\varphi}} \right), \quad p(\vartheta) = \frac{1}{\sqrt{2\pi\Sigma_{\vartheta}}} \exp\left(-\frac{\vartheta - \bar{\vartheta}}{2\Sigma_{\vartheta}} \right), \tag{9.5}$$

where, Σ_{φ} and Σ_{ϑ} are variances. From this, the joint probability is given by the following equation:

$$\begin{aligned} p(\boldsymbol{\varphi}, \vartheta) &= p(\boldsymbol{\varphi} \mid \vartheta)\, p(\boldsymbol{\varphi}, \vartheta) \\ &= \prod_{i=1}^{N} \frac{1}{\sqrt{2\pi\Sigma_{\varphi}}} \exp\left(-\frac{(\varphi_i - \vartheta)^2}{2\Sigma_{\varphi}} \right) \cdot \frac{1}{\sqrt{2\pi\Sigma_{\vartheta}}} \exp\left(-\frac{(\vartheta - \bar{\vartheta})^2}{2\Sigma_{\vartheta}} \right). \end{aligned} \tag{9.6}$$

In the case of the maximum likelihood method, the natural logarithm is taken and inferred as follows:

$$\vartheta = \arg\max_{\vartheta} \ln p(\boldsymbol{\varphi} \mid \vartheta) = \arg\max_{\vartheta} \left[-\frac{1}{2} \ln(2\pi\Sigma_{\varphi}) - \sum_{i=1}^{N} \frac{(\varphi_i - \vartheta)^2}{2\Sigma_{\varphi}} \right] = \frac{1}{N} \sum_{i=1}^{N} \varphi_i \,. \tag{9.7}$$

The variance σ_z may also be an unknown. In the maximum likelihood method, as shown in equation (9.7), the value that maximizes the likelihood is used as the estimated value of ϑ, but this is not well-founded. However, if the prior probability is constant regardless of ϑ in Bayesian inference, the maximum likelihood method and Bayesian inference coincide.

Bayesian inference uses Bayes' theorem:

$$p(\vartheta \mid \boldsymbol{\varphi}) = \frac{p(\boldsymbol{\varphi}, \vartheta)}{p(\boldsymbol{\varphi})} = \frac{p(\boldsymbol{\varphi} \mid \vartheta)\, p(\vartheta)}{p(\boldsymbol{\varphi})} \,. \tag{9.8}$$

In the MAP (Maximum a posteriori) estimation, the denominator $p(\boldsymbol{\varphi})$ is not involved in the maximum value search, so consider the logarithm of the numerator:

$$\begin{aligned} \vartheta &= \arg\max_{\vartheta} \ln p(\boldsymbol{\varphi} \mid \vartheta)\, p(\vartheta) = \arg\max_{\vartheta} (\ln p(\boldsymbol{\varphi}, \vartheta)) \\ &= -\arg\max_{\vartheta} \left\{ \sum_{i=1}^{N} \left[\frac{1}{2} \ln(2\pi\Sigma_{\varphi}) + \frac{(\varphi_i - \vartheta)^2}{2\Sigma_{\varphi}} \right] + \frac{1}{2} \ln(2\pi\Sigma_{\vartheta}) + \frac{(\vartheta - \bar{\vartheta})^2}{2\Sigma_{\vartheta}} \right\}. \end{aligned} \tag{9.9}$$

There is good evidence for this estimation method. In-vivo inference on the right side of Fig. 9.1 can also be considered as Bayesian reasoning. The following equation is obtained from this equation:

$$\frac{\partial \ln p(\boldsymbol{\varphi},\vartheta)}{\partial \vartheta}=\frac{1}{\Sigma_\varphi}\sum_{i=1}^{N}(\varphi_i-\vartheta)-\frac{1}{\Sigma_\vartheta}(\vartheta-\bar{\vartheta})=0, \tag{9.10}$$

The following equation can be obtained by solving this:

$$\vartheta=\frac{\Sigma_\vartheta}{\Sigma_\varphi+N\Sigma_\vartheta}\sum_{i=1}^{N}\varphi_i+\frac{\Sigma_\varphi}{\Sigma_\varphi+N\Sigma_\vartheta}\bar{\vartheta}. \tag{9.11}$$

When the variance Σ_ϑ is large (that is, the accuracy of φ is low or the accuracy of $\bar{\vartheta}$ is high), ϑ becom $\vartheta=(1/N)\Sigma_{i=1}^{N}\varphi_i$. And, when the variance Σ_φ is large (that is, the accuracy of φ is low or the accuracy of $\bar{\vartheta}$ is high), it becomes $\vartheta=\bar{\vartheta}$.

The likelihood function is formed by learning. That is, in the case of measurement by the camera on the left side of Fig. 9.1, the frequency distribution is formed by repeated imaging of the number of pixels of the camera image. Similarly, in the case of eye measurement by the person on the right side, it is the formation of the frequency distribution of the number of pixels of the retinal image by repeated observation. Furthermore, if it is $N\to\infty$, it becomes $\vartheta=(1/N)\Sigma_{i=1}^{N}\varphi_i$. It can be seen that the accuracy improves as the number of observations increases.

If $\bar{\vartheta}$ is also unknown, then equation (9.10) is replaced by:

$$\begin{aligned}\frac{\partial \ln p(\boldsymbol{\varphi},\vartheta)}{\partial \vartheta}&=\frac{1}{\Sigma_\varphi}\sum_{i=1}^{N}(\varphi_i-\vartheta)-\frac{1}{\Sigma_\vartheta}(\vartheta-\bar{\vartheta})=0,\\ \frac{\partial \ln p(\boldsymbol{\varphi},\vartheta)}{\partial \bar{\vartheta}}&=\frac{1}{\Sigma_\vartheta}(\vartheta-\bar{\vartheta})=0.\end{aligned} \tag{9.12}$$

Solving this, we get the following equation:

$$\vartheta=\bar{\vartheta}=\frac{1}{N}\sum_{i=1}^{N}\varphi_i. \tag{9.13}$$

In the above discussion, the variances Σ_φ and Σ_ϑ may also be unknowns. Since the partial differential of $\ln p(\boldsymbol{\varphi},\vartheta)$ by Σ_φ and Σ_ϑ is the reciprocal of these, it is considered that learning by the steepest descent method, which is considered to be feasible even in the living body, is slow in the initial state where these are large. It can be considered to correspond to long-term learning.

9.2.3 Derivation of variational free energy

Based on the above introduction, let us consider the information processing in the brain by taking the case of Fig. 9.1 as an example. The length of the bar is estimated from the camera image on the left side of the figure and from the retinal image on the right side. The recognition of the length of the bar is ϑ, and the sensation or retinal image is φ. Let's think firstly Kullback-Leibler divergence (KLD) of $q(\vartheta)$ and $p(\vartheta|\varphi)$. $p(\vartheta|\varphi)$ is the posterior probability, and $q(\vartheta)$ is the approximate solution of the posterior probability:

$$KL(q(\vartheta)\,||\,p(\vartheta|\varphi)) = \int q(\vartheta)\ln\frac{q(\vartheta)}{p(\vartheta|\varphi)}d\vartheta \tag{9.14}$$

Considering the non-negativeness of KLD, KLD is minimized when $q(\vartheta)=p(\vartheta|\varphi)$. However, even if equation (9.13) itself is used as the minimum value problem for finding $p(\vartheta|\varphi)$, a meaningful answer cannot be found. Variational Free Energy (VFE) can be obtained by rewriting equation (9.13). Firstly, we have

$$\begin{aligned} KL(q(\vartheta)\,||\,p(\vartheta|\varphi)) &= \int q(\vartheta)\log\frac{q(\vartheta)p(\varphi)}{p(\vartheta,\varphi)}d\vartheta \\ &= \int q(\vartheta)\log\frac{q(\vartheta)}{p(\vartheta,\varphi)}d\vartheta + \int q(\vartheta)\log p(\varphi)d\vartheta \\ &= \int q(\vartheta)\log\frac{q(\vartheta)}{p(\vartheta,\varphi)}d\vartheta + \log p(\varphi) = KL(q(\vartheta)\,||\,p(\vartheta,\varphi)) + \log p(\varphi), \end{aligned} \tag{9.15}$$

and rewrting further, we finally derive the variational free energy F:

$$\begin{aligned} F \equiv KL(q(\vartheta)\,||\,p(\vartheta,\varphi)) &= \int q(\vartheta)\ln\frac{q(\vartheta)}{p(\vartheta,\varphi)}d\vartheta \\ &= KL(q(\vartheta)\,||\,p(\vartheta|\varphi)) - \ln p(\vartheta) \end{aligned} . \tag{9.16}$$

At first glance, the unknown variable of the minimum value problem of F seems to be only recognition ϑ, but the sense φ is also an unknown variable. The second term on the rightest is Shannon Surprise. Since both of the two terms on the right-hand side of equation (9.16) are non-negative (Appendix 9A), q that minimizes F is $p(\vartheta|\varphi)$, but it can be seen that F also has the effect of minimizing surprises. Since the sensation and the physical movement (action) are directly connected, the action causes a change in the sensation φ, which improves the accuracy of reasoning. Improving the accuracy of reasoning through actions is called active reasoning. Active reasoning was first clearly demonstrated by the principle of minimal free energy by Friston [1].

9.2.4 Various meaning of variational free energy

Equation (9.16) is the free energy proposed by Friston [1], but a probability distribution other than the original probability distribution is introduced. If $q(\vartheta)$ is freely taken and the minimum condition of F defined by equation (9.16) is obtained, $q(\vartheta)$ becomes $p(\vartheta|\varphi)$. Therefore, equation (9.16) is called variational free energy.

On the other hand, if F is minimized with the degrees of freedom of $q(\vartheta)$ constrained, an approximate solution of $q(\vartheta)$ can be obtained, which is considered to be an approximation of $p(\vartheta|\varphi)$. This is the method for finding the approximation of $p(\vartheta|\varphi)$.

When writing F in the form of equation (9.15), the following interpretation is given:

$$\begin{aligned} F &\equiv KL(q(\vartheta)\,\|\,p(\vartheta,\varphi)) = KL(q(\vartheta)\,\|\,p(\vartheta|\varphi)) - \log p(\varphi) \\ &= \text{Relative Entropy} + \text{Sensory Surprisal (Shanon surprise)}. \end{aligned} \tag{9.17}$$

Rewriting F also gives the following interpretation of F:

$$\begin{aligned} F &\equiv KL(q(\vartheta)\,\|\,p(\vartheta,\varphi)) = \int q(\vartheta)\log\frac{q(\vartheta)}{p(\vartheta,\varphi)}d\vartheta \\ &= \int q(\vartheta)\log\frac{q(\vartheta)}{p(\varphi|\vartheta)p(\vartheta)}d\vartheta \\ &= -\int q(\vartheta)\log p(\varphi|\vartheta)d\vartheta + \int q(\vartheta)\log\frac{q(\vartheta)}{p(\vartheta)}d\vartheta \\ &= -\int q(\vartheta)\log p(\varphi|\vartheta)dv + KL(q(\vartheta)\,\|\,p(\vartheta)) \\ &= \text{Uncertainty} + \text{Bayesian Surprise (Complexity)}. \end{aligned} \tag{9.18}$$

Thinking in the same way, there are also the following interpretations:

$$\begin{aligned} F &\equiv KL(q(\vartheta)\,\|\,p(\vartheta,\varphi)) = -\int q(\vartheta)\log p(\varphi,\vartheta)dv + \int q(\vartheta)\log q(\vartheta)d\vartheta \\ &= -Q\text{-Funtion} - \text{Entropy}. \end{aligned} \tag{9.19}$$

9.2.5 Approximation of variational free energy

(1) The simplest approximation (estimation of the mean value of the posterior distribution)

Replacing Bayesian inference with the principle of minimum variational free energy, and using Gaussian distribution approximation and the steepest descent method, can lead to calculations feasible in the brain. For simplicity, we consider the case of one variable. First, we consider a Gaussian approximation of the likelihood function and prior distribution:

$$p(\varphi|\vartheta) = \frac{1}{\sqrt{2\pi\Sigma_\varphi}}\exp\left(-\frac{(\varphi - g(\vartheta))^2}{2\Sigma_\varphi}\right).\quad p(\vartheta) = \frac{1}{\sqrt{2\pi\Sigma_\vartheta}}\exp\left(-\frac{(\vartheta - \bar{\mu})^2}{2\Sigma_{\bar{\vartheta}}}\right). \tag{9.20}$$

From this, in the variational free energy of equation (9.16), the joint distribution is approximated as follows:

$$p(\varphi,\vartheta) = \frac{1}{\sqrt{2\pi\Sigma_\varphi}}\frac{1}{\sqrt{2\pi\Sigma_\vartheta}}\exp\left(-\frac{(\varphi - g(\vartheta))^2}{2\Sigma_\varphi} - \frac{(\vartheta - \bar{\mu})^2}{2\Sigma_{\bar{\vartheta}}}\right). \tag{9.21}$$

In §9.2.4, it was stated that an approximation of the posterior distribution $p(\vartheta|\varphi)$ can be obtained by approximating $q(\vartheta)$ under the condition of minimum variational free energy. Therefore, let us assume q in the following form, which means a normal distribution with a mean μ variance of 0:

$$q(\vartheta)=\delta(\vartheta-\mu), \tag{9.22}$$

where, $\delta(\vartheta)$ is a Dirac delta function. The unknown variable is μ. Substituting equations (9.21) and (9.22) into equation (9.16) yields the following equation:

$$\begin{aligned}F[\mu]&=\int\delta(\vartheta-\bar{\mu})\ln\delta(\vartheta-\bar{\mu})-\int\delta(\vartheta-\bar{\mu})\ln p(\vartheta,\varphi)\\&=\ln\delta(0)+\frac{1}{2}\int\delta(\vartheta-\mu)\left(\frac{(\varphi-g(\vartheta))^2}{\Sigma_\varphi}+\frac{(\vartheta-\bar{\mu})^2}{\Sigma_\vartheta}+\ln 4\pi^2\sqrt{\Sigma_\varphi\Sigma_\vartheta}\right)d\vartheta\\&\sim\frac{1}{2}\left(\frac{(\varphi-g(\mu))^2}{\Sigma_\varphi}+\frac{(\mu-\bar{\mu})^2}{\Sigma_\vartheta}+\ln 4\pi^2\sqrt{\Sigma_\varphi\Sigma_\vartheta}\right).\end{aligned} \tag{9.23}$$

The term $\ln\delta(0)$ is a constant term and can be omitted. The integral for ϑ is done in $-\infty<\vartheta<\infty$. Also, $F[\mu]$ is used to indicate that μ is an unknown variable. Therefore, the principle of variational free energy minimum should be simplified and the following minimum value problem should be solved:

$$\begin{aligned}\mu&=\underset{\mu}{\arg\min}\left[\ln p(\mu,\varphi)\right]\\&=\underset{\mu}{\arg\min}\frac{1}{2}\left(\frac{(\varphi-g(\mu))^2}{\Sigma_\varphi}+\frac{(\mu-\bar{\mu})^2}{\Sigma_\vartheta}+\log 4\pi^2\sqrt{\Sigma_\varphi\Sigma_\vartheta}\right).\end{aligned} \tag{9.24}$$

What should not be forgotten here is that the effect of the action on improving the accuracy of the senses is not forgotten. Equation (9.24) looks like this if it also includes actions:

$$\begin{aligned}(\mu,a)&=\underset{(\mu,a)}{\arg\min}F=\underset{(\mu,a)}{\arg\min}\left[\ln p(\mu,\varphi)\right]\\&=\underset{(\mu,a)}{\arg\min}\frac{1}{2}\left(\frac{(\varphi-g(\mu))^2}{\Sigma_\varphi}+\frac{(\mu-\bar{\mu})^2}{\Sigma_\vartheta}+\ln 4\pi^2\sqrt{\Sigma_\varphi\Sigma_\vartheta}\right).\end{aligned} \tag{9.25}$$

To solve equation (9.25) by the steepest descent method, it becomes:

$$(\mu,a)_{t+\delta t}=(\mu,a)_t+\left(\frac{\partial F}{\partial\mu},\frac{\partial F}{\partial\varphi}\frac{d\varphi}{da}\right)\cdot k\delta t. \tag{9.26}$$

Here, t is a sequential step, δt is a sequential unit, and k is a sequential parameter. Since the algorithms of equations (9.25) and (9.26) are simple, it is considered that they can be implemented in vivo.

(2) A second simplest approximation (estimation of mean value μ and variance ζ of posterior distribution ... with the assumption of a sharp peak in joint distribution)

We consider the Gaussian approximation for the likelihood function $p(\varphi|\vartheta)$ and the prior distribution $p(\vartheta)$ remains in equation (9.20), so the joint distribution remains $p(\vartheta|\varphi)$ in equation (9.21). And $q(\vartheta)$, which is an approximation of the posterior distribution $p(\vartheta|\varphi)$, is replaced by equation (9.22) with mean μ and variance ζ. Let us consider the case of approximation by the Gaussian distribution [2]. That is,

$$p(\varphi,\vartheta)=\frac{1}{\sqrt{2\pi\Sigma_\varphi}}\frac{1}{\sqrt{2\pi\Sigma_\vartheta}}\exp\left(-\frac{(\varphi-g(\vartheta))^2}{2\Sigma_\varphi}-\frac{(\vartheta-\bar{\mu})^2}{2\Sigma_\vartheta}\right) \tag{9.27}$$

and

$$q(\vartheta)=\frac{1}{\sqrt{2\pi\zeta}}\exp\left(-\frac{(\vartheta-\mu)^2}{2\zeta}\right). \tag{9.28}$$

This time, μ and ζ are unknown variables.

Substituting equations (9.27) and (9.28) for the variational free energy F gives the following equation:

$$\begin{aligned}F[\mu,\zeta]&=\int q(\vartheta)\ln\frac{q(\vartheta)}{p(\vartheta,\varphi)}d\vartheta\\&=\int q(\vartheta)\left(-\ln\sqrt{2\pi\zeta}-\frac{(\vartheta-\mu)^2}{2\zeta}\right)d\vartheta+\int q(\vartheta)E(\vartheta,\varphi)d\vartheta\\&=-\ln\sqrt{2\pi\zeta}-\frac{1}{2}+\int q(\vartheta)E(\vartheta,\varphi)d\vartheta,\end{aligned} \tag{9.29}$$

where

$$E(\vartheta,\varphi)=-\ln p(\vartheta,\varphi)=\frac{1}{2}\left(\ln(\Sigma_\varphi\Sigma_\vartheta)+\frac{(\varphi-g(\vartheta))^2}{\Sigma_\varphi}+\frac{(\vartheta-\bar{\mu})^2}{\Sigma_\vartheta}\right)+\ln(2\pi). \tag{9.30}$$

$E(\vartheta,\varphi)$ is called energy. In the following discussion, the last constant on the right side is omitted.

Now, for the sake of simple results, assume that $E(\vartheta,\varphi)$ has a sharp peak at $\vartheta=\mu$:

$$E(\vartheta,\varphi)\approx E(\mu,\varphi)+\left[\frac{\partial E}{\partial\vartheta}\right]_{\vartheta=\mu}(\vartheta-\mu)+\frac{1}{2}\left[\frac{\partial^2 E}{\partial\vartheta^2}\right]_{\vartheta=\mu}(\vartheta-\mu)^2. \tag{9.31}$$

Approximating the last term on the right-hand side of equation (9.29) using equation (9B.5), we obtain

$$\begin{aligned}\int q(\vartheta)E(\vartheta,\varphi)d\vartheta &\approx E(\mu,\varphi)+\frac{1}{2}\left[\frac{\partial^2 E}{\partial \vartheta^2}\right]_{\vartheta=\mu}\int q(\vartheta)(\vartheta-\mu)^2 d\vartheta \\ &= E(\mu,\varphi)+\frac{1}{2}\left[\frac{\partial^2 E}{\partial \vartheta^2}\right]_{\vartheta=\mu}\zeta.\end{aligned} \tag{9.32}$$

Substituting equation (9B.6) into equation (9B.3) yields the following equation:

$$\begin{aligned}F[\mu,\zeta] &= -\ln\sqrt{2\pi\zeta}-\frac{1}{2}+E(\mu,\varphi)+\frac{1}{2}\left[\frac{\partial^2 E}{\partial \vartheta^2}\right]_{\vartheta=\mu}\zeta \\ &= E(\mu,\varphi)+\frac{1}{2}\left(\left[\frac{\partial^2 E}{\partial \vartheta^2}\right]_{\vartheta=\mu}\zeta-\ln(2\pi\zeta)-1\right).\end{aligned} \tag{9.33}$$

To minimize F, the derivative of F by ζ should also be 0, so let ζ^* be ζ that satisfies this condition, then ζ^* is:

$$\left[\frac{\partial^2 E}{\partial \vartheta^2}\right]_{\vartheta=\mu}-\frac{1}{\zeta^*}=0 \;\rightarrow\; \zeta^*=\left(\left[\frac{\partial^2 E}{\partial \vartheta^2}\right]_{\vartheta=\mu}\right)^{-1}. \tag{9.34}$$

Substituting equation (9.34) into equation (9.33) yields the following equation:

$$F=E(\mu,\varphi)-\frac{1}{2}\ln(2\pi\zeta^*). \tag{9.35}$$

In the calculation of the mean μ, the second term on the right side can be omitted in the calculation of equation (9.35):

$$F[\mu]=E(\mu,\varphi). \tag{9.36}$$

After all, the mean of $q(\vartheta)$, which is an approximation of the posterior distribution $p(\vartheta|\varphi)$, can be calculated by minimizing equation (9.36), and the variance can be calculated by equation (9.34).

(1) The most accurate approximation (estimation of mean value μ and variance ζ of posterior distribution ... no assumption of sharp peak in joint distribution)

Without assuming the assumption of equation (9.31) that the joint distribution $p(\vartheta,\varphi)$ has a sharp peak, we consider the variational free energy F without introducing any approximation other than the Gaussian approximation:

$$
\begin{aligned}
F[\mu,\zeta] &= \int q(\vartheta)\ln\frac{q(\vartheta)}{p(\vartheta,\varphi)}\,d\vartheta \\
&= \int \frac{1}{\sqrt{2\pi\zeta}}\,exp\left(-\frac{(\vartheta-\mu)^2}{2\zeta}\right)\left(\begin{array}{l} -\ln\sqrt{2\pi\zeta}-\dfrac{(\vartheta-\mu)^2}{2\zeta} \\ +\ln(2\pi)+\dfrac{1}{2}\ln(\Sigma_\varphi\Sigma_\vartheta) \\ +\dfrac{(\varphi-g(\vartheta))^2}{2\Sigma_\varphi}+\dfrac{(\vartheta-\bar{\mu})^2}{2\Sigma_\vartheta} \end{array}\right)d\vartheta.
\end{aligned}
\tag{9.37}
$$

If we consider the minimum value problem by the steepest descent method, the mean μ and variance ζ of the posterior distribution $p(\vartheta|\varphi)$ are the solutions to the following problem:

$$
(\mu,\zeta)=\underset{(\mu,\zeta)}{\arg\min}\,F(\mu,\zeta). \tag{9.38}
$$

Specifically, it becomes the following equation:

$$
(\mu,\zeta)_{t+\delta t}=(\mu,\zeta)_{t+\delta t}+\left(\frac{\partial F}{\partial\mu},\frac{\partial F}{\partial\zeta}\right)_t k\delta t. \tag{9.39}
$$

As can be seen from equation (9.37), $\partial F/\partial\mu$ and $\partial F/\partial\zeta$ appear to be too complex to be implemented in vivo.

We consider a calculation that uses the variational free energy itself given by equation. (9.37) from a mathematical interest. In preparation for that, the posterior distribution $p(\vartheta|\varphi)$ is derived from the joint distribution $p(\varphi,\vartheta)$:

$$
p(\varphi,\vartheta)=\frac{1}{\sqrt{2\pi\Sigma_\varphi}}\frac{1}{\sqrt{2\pi\Sigma_\vartheta}}\exp\left(-\frac{(\varphi-g(\vartheta))^2}{2\Sigma_\varphi}-\frac{(\vartheta-\bar{\mu})^2}{2\Sigma_\vartheta}\right). \tag{9.21}
$$

Integrating $p(\varphi,\vartheta)$ with ϑ gives the following equation:

$$
\begin{aligned}
p(\varphi) &= \int_{-\infty}^{\infty} p(\varphi,\vartheta)\,d\vartheta \\
&= \frac{1}{\sqrt{2\pi\Sigma_\varphi}}\frac{1}{\sqrt{2\pi\Sigma_\vartheta}}\int_{-\infty}^{\infty}\exp\left(-\frac{(\varphi-g(\vartheta))^2}{2\Sigma_\varphi}-\frac{(\vartheta-\bar{\mu})^2}{2\Sigma_\vartheta}\right)d\vartheta.
\end{aligned}
\tag{9.40}
$$

Using this, $p(\vartheta|\varphi)$ can be found as follows:

$$p(\vartheta|\varphi)=\frac{p(\vartheta,\varphi)}{p(\varphi)}=\frac{\exp\left(-\frac{(\varphi-g(\vartheta))^2}{2\Sigma_\varphi}-\frac{(\vartheta-\bar{\mu})^2}{2\Sigma_\vartheta}\right)}{\int_{-\infty}^{\infty}\exp\left(-\frac{(\varphi-g(\vartheta))^2}{2\Sigma_\varphi}-\frac{(\vartheta-\bar{\mu})^2}{2\Sigma_\vartheta}\right)d\vartheta}. \tag{9.41}$$

In the case of linear, that is, $g(\vartheta)=\vartheta$, equations (9.40) and (9.41) are as follows:

$$\begin{aligned}p(\varphi)&=\frac{1}{\sqrt{(2\pi)^2\Sigma_\varphi\Sigma_\vartheta}}\sqrt{2\pi\frac{\Sigma_\varphi\Sigma_\vartheta}{\Sigma_\vartheta+\Sigma_\varphi}}\exp\left(-\frac{(\bar{\mu}-\varphi)^2}{2(\Sigma_\vartheta+\Sigma_\varphi)}\right)\\&=\frac{1}{\sqrt{2\pi(\Sigma_\vartheta+\Sigma_\varphi)}}\exp\left(-\frac{(\varphi-\bar{\mu})^2}{2(\Sigma_\vartheta+\Sigma_\varphi)}\right)\end{aligned}, \tag{9.42}$$

$$\begin{aligned}&p(\vartheta|\varphi)\\&=\frac{\exp\left(-\frac{(\varphi-\vartheta)^2}{2\Sigma_\varphi}-\frac{(\vartheta-\bar{\mu})^2}{2\Sigma_\vartheta}\right)}{\int_{-\infty}^{\infty}\exp\left(-\frac{(\varphi-\vartheta)^2}{2\Sigma_\varphi}-\frac{(\vartheta-\bar{\mu})^2}{2\Sigma_\vartheta}\right)d\vartheta}=\sqrt{\frac{\Sigma_\vartheta+\Sigma_\varphi}{2\pi\Sigma_\varphi\Sigma_\vartheta}}\exp\left(-\frac{\left(\vartheta-\frac{\Sigma_\varphi\bar{\mu}+\Sigma_\vartheta\varphi}{\Sigma_\vartheta+\Sigma_\varphi}\right)^2}{\frac{2\Sigma_\varphi\Sigma_\vartheta}{\Sigma_\vartheta+\Sigma_\varphi}}\right).\end{aligned} \tag{9.43}$$

This is the correct answer for posterior probabilities $p(\vartheta\,|\varphi)$ when the joint probabilities is given by equation (9.27) and $g(\vartheta)=\vartheta$. The mean μ and variance ζ are given by:

$$\mu=\frac{\Sigma_\varphi\bar{\mu}+\Sigma_\vartheta\varphi}{\Sigma_\vartheta+\Sigma_\varphi},\quad \zeta=\frac{\Sigma_\varphi\Sigma_\vartheta}{\Sigma_\vartheta+\Sigma_\varphi}. \tag{9.44}$$

Next, consider a method for calculating $p(\vartheta|\varphi)$ from equation (9.37). first, we use

$$\begin{aligned}&\int_{-\infty}^{\infty}exp\left(-\frac{(\vartheta-\mu)^2}{2\zeta}\right)d\vartheta=\sqrt{2\zeta\pi},\\&\int_{-\infty}^{\infty}(\vartheta-\mu)^2exp\left(-\frac{(\vartheta-\mu)^2}{2\zeta}\right)d\vartheta=(2\zeta)^{3/2}\Gamma\left(\frac{3}{2}\right)=(2\zeta)^{3/2}\frac{\sqrt{\pi}}{2},\end{aligned} \tag{9.45}$$

and simplify equation (9.37):

$$\begin{aligned}F(\mu,\zeta)&=\left(-\frac{1}{2}\ln\zeta+\frac{1}{2}\ln(\Sigma_\varphi\Sigma_\vartheta)+\frac{1}{2}\ln(2\pi)+\frac{(\mu-\bar{\mu})^2}{2\Sigma_\vartheta}\right)\\&+\zeta\left(-\frac{1}{2\zeta}+\frac{1}{2\Sigma_\vartheta}\right)+\frac{1}{\sqrt{2\pi\zeta}}\int_{-\infty}^{\infty}exp\left(-\frac{(\vartheta-\mu)^2}{2\zeta}\right)\frac{(\varphi-g(\vartheta))^2}{2\Sigma_\varphi}d\vartheta.\end{aligned} \tag{9.46}$$

From equation (9.46), $M = \partial F/\partial\mu = 0$ and $Z = \partial F/\partial\zeta = 0$ are given by the following equation using numerical differentiation:

$$M(\mu,\zeta) \equiv \frac{\partial F(\mu,\zeta)}{\partial\mu} = \frac{F(\mu+d\mu,\zeta)-F(\mu,\zeta)}{d\mu} = 0, \tag{9.47a}$$

$$Z(\mu,\zeta) \equiv \frac{\partial F(\mu,\zeta)}{\partial\zeta} = \frac{F(\mu,\zeta+d\zeta)-F(\mu,\zeta)}{d\zeta} = 0. \tag{9.47b}$$

Using Newton-Raphson's method to solve Eq. (9.47) gives a two-dimensional simultaneous algebraic equation for $d\mu$ and $d\zeta$:

$$\frac{\partial M}{\partial\mu}d\mu + \frac{\partial M}{\partial\zeta}d\zeta = -M, \tag{9.48a}$$

$$\frac{\partial Z}{\partial\mu}d\mu + \frac{\partial Z}{\partial\zeta}d\zeta = -Z. \tag{9.48b}$$

The derivative of M and Z is also calculated by numerical differentiation. Solving this, we can find $d\mu$ and $d\zeta$:

$$d\mu = -\frac{M\dfrac{\partial Z}{\partial\zeta} - Z\dfrac{\partial M}{\partial\zeta}}{\dfrac{\partial M}{\partial\mu}\dfrac{\partial Z}{\partial\zeta} - \dfrac{\partial Z}{\partial\mu}\dfrac{\partial M}{\partial\zeta}}, \quad d\zeta = -\frac{\dfrac{\partial M}{\partial\mu}Z - \dfrac{\partial Z}{\partial\mu}M}{\dfrac{\partial M}{\partial\mu}\dfrac{\partial Z}{\partial\zeta} - \dfrac{\partial Z}{\partial\mu}\dfrac{\partial M}{\partial\zeta}}. \tag{9.49}$$

Therefore, the convergence value of the iterative calculation that updates μ and ζ according to the following equation is the solution:

$$\mu_{new} = \mu_{old} + d\mu, \quad \zeta_{new} = \zeta_{old} + d\zeta \tag{9.50}$$

By substituting this into equation (9.37), the variational free energy F can be obtained.

The calculation code in C language is shown in Appendix 9B. Numerical calculations were actually performed, and the result by equation (9.43) and the result by the minimum variational free energy were compared. The following equation was assumed in the calculation:

$$g(\vartheta) = \vartheta. \tag{9.51}$$

The parameter values used in the calculation are shown below:

$$\bar{\mu} = 1, \quad \Sigma_\varphi = 1.5, \quad \Sigma_\vartheta = 2.5, \quad d\mu = d\zeta = 0.000001. \tag{9.52}$$

In Fig. 9.3, the initial values of μ and ζ are set only for the minimum value of φ, and φ is sequentially increased while the previous convergence value is used as the initial value of the next calculation. For the initial value at the start, refer to equation

(9.44). Figures 9.3 and 9.4 show a comparison between the correct answer given in equations (9.44) and (9.43) and the approximate solution obtained by solving equation (9.47), and sufficient accuracy is obtained.

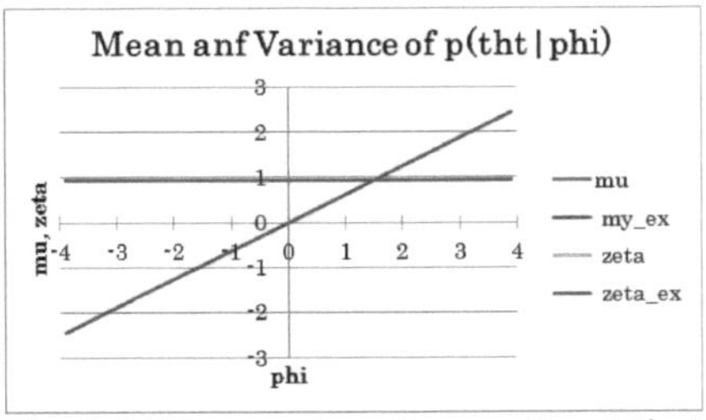

Fig. 9.3 Mean and variance of $p(\vartheta|\varphi)$

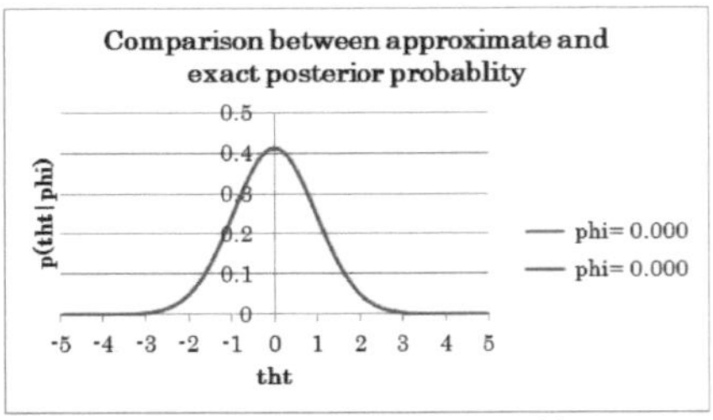

Fig. 9.4 Posterior probability $p(\vartheta|\varphi)$ at $\varphi=0$

9.2.5 Variational free energy of multi-variables

Rewriting the one-dimensional equation (9.23), the variational free energy minimum principle becomes a minimum problem of energy $E=-\ln p(\vartheta,\varphi)=-\ln p(\mu,\varphi)$:

$$E(\varphi,\mu)=\frac{1}{2}\left(\frac{\varepsilon_\varphi{}^2}{\Sigma_\varphi}+\frac{\varepsilon_\vartheta{}^2}{\Sigma_\vartheta}+\ln(\Sigma_\varphi\Sigma_\vartheta)\right),$$
$$\varepsilon_\varphi=\varphi-g(\mu),\quad \varepsilon_\vartheta=\mu-\bar{\mu}. \tag{9.53}$$

The constant term is omitted in equation (9.52). Considering active reasoning, the unknown variables of E are φ and μ. The likelihood function and prior probability is then given by the following equations:

$$p(\varphi|\mu)=\frac{1}{\sqrt{(2\pi)^N\Sigma_\varphi}}\exp\left(-\frac{1}{2}(\varphi-g(\mu))\Sigma_\varphi{}^{-1}(\varphi-g(\mu))\right), \tag{9.54}$$

$$p(\mu)=\frac{1}{\sqrt{(2\pi)^N\Sigma_\vartheta}}\exp\left(-\frac{1}{2}(\mu-\bar{\mu})\Sigma_\vartheta(\mu-\bar{\mu})^T\right), \tag{9.55}$$

Let us consider an extension to multiple dimensions. Suppose a one-dimensional variable $\varphi, \mu, \bar{\mu}$ is expanded to N dimensions to become $\boldsymbol{\varphi}, \boldsymbol{\mu}, \bar{\boldsymbol{\mu}}$:

$$\boldsymbol{\varphi} = (\varphi_1, \varphi_2, \cdots, \varphi_N)^T, \quad \boldsymbol{\mu} = (\mu_1, \mu_2, \cdots, \mu_N)^T, \quad \bar{\boldsymbol{\mu}} = (\bar{\mu}_1, \bar{\mu}_2, \cdots, \bar{\mu}_N)^T. \tag{9.56}$$

At this time, the likelihood function, prior probability, and energy are each extended as follows:

$$p(\boldsymbol{\varphi} \mid \boldsymbol{\mu}) = \frac{1}{\sqrt{(2\pi)^N \mid \boldsymbol{\Sigma}_\varphi \mid}} \exp\left(-\frac{1}{2}(\boldsymbol{\varphi} - g(\boldsymbol{\mu}))\boldsymbol{\Sigma}_\varphi^{-1}(\boldsymbol{\varphi} - g(\boldsymbol{\mu}))^T \right), \tag{9.57}$$

$$p(\boldsymbol{\mu}) = \frac{1}{\sqrt{(2\pi)^N \mid \boldsymbol{\Sigma}_\vartheta \mid}} \exp\left(-\frac{1}{2}(\boldsymbol{\mu} - \bar{\boldsymbol{\mu}})\boldsymbol{\Sigma}_\vartheta^{-1}(\boldsymbol{\mu} - \bar{\boldsymbol{\mu}})^T \right), \tag{9.58}$$

$$\begin{aligned} E(\boldsymbol{\mu}, \boldsymbol{\varphi}) &= -\ln p(\boldsymbol{\mu}, \boldsymbol{\varphi}) = -\ln p(\boldsymbol{\varphi} \mid \boldsymbol{\mu}) - \ln p(\boldsymbol{\mu}) \\ &= \frac{1}{2}(\boldsymbol{\varphi} - g(\boldsymbol{\mu}))\boldsymbol{\Sigma}_\varphi^{-1}(\boldsymbol{\varphi} - g(\boldsymbol{\mu}))^T + \frac{1}{2}\ln \mid \boldsymbol{\Sigma}_\varphi \mid + \frac{1}{2}(\boldsymbol{\mu} - \bar{\boldsymbol{\mu}})\boldsymbol{\Sigma}_\vartheta^{-1}(\boldsymbol{\mu} - \bar{\boldsymbol{\mu}}))^T + \frac{1}{2}\ln \mid \boldsymbol{\Sigma}_\vartheta \mid. \end{aligned} \tag{9.59}$$

Here, $\boldsymbol{\Sigma}_\vartheta$ and $\boldsymbol{\Sigma}_\varphi$ are covariance matrices. $\boldsymbol{\Sigma}_\vartheta^{-1}$ and $\boldsymbol{\Sigma}_\varphi^{-1}$ are inverse matrices, and $|\boldsymbol{\Sigma}_\vartheta|$ and $|\boldsymbol{\Sigma}_\varphi|$ are determinants.

In this paper, we assume the independence of cognitive and sensory variables. Then, the likelihood function and prior probabilities can be simplified as follows:

$$\begin{aligned} p(\boldsymbol{\varphi} \mid \boldsymbol{\mu}) &= p(\varphi_1, \varphi_2, \cdots, \varphi_N \mid \mu_1, \mu_2, \cdots, \mu_N) \\ &= \prod_{\alpha=1}^{N} p(\varphi_\alpha \mid \mu_\alpha) = \prod_{\alpha=1}^{N} \frac{1}{\sqrt{2\pi\Sigma_\varphi^\alpha}} \exp\left(-\frac{(\varphi_\alpha - g(\mu_\alpha))^2}{2\Sigma_\varphi^\alpha} \right), \end{aligned} \tag{9.60}$$

$$p(\boldsymbol{\mu}) = p(\mu_1, \mu_2, \cdots, \mu_N) = \prod_{\alpha=1}^{N} p(\mu_\alpha) = \prod_{\alpha=1}^{N} \frac{1}{\sqrt{2\pi\Sigma_\vartheta^\alpha}} \exp\left(-\frac{(\mu_\alpha - \bar{\mu}_\alpha)^2}{2\Sigma_\vartheta^\alpha} \right), \tag{9.61}$$

Therefore, the energy is given by:

$$\begin{aligned} E(\boldsymbol{\mu}, \boldsymbol{\varphi}) &= -\ln p(\boldsymbol{\mu}, \boldsymbol{\varphi}) \\ &= -\ln p(\boldsymbol{\varphi} \mid \boldsymbol{\mu}) - \ln p(\boldsymbol{\mu}) = \frac{1}{2}\sum_{\alpha=1}^{N}\left(\frac{\varepsilon_\varphi^{\alpha 2}}{\Sigma_\varphi^\alpha} + \frac{\varepsilon_\vartheta^{\alpha 2}}{\Sigma_\vartheta^\alpha} + \ln(\Sigma_\varphi^\alpha \Sigma_\vartheta^\alpha) \right), \end{aligned} \tag{9.62}$$

$$\varepsilon_\varphi^\alpha = \varphi_\alpha - g(\mu\alpha), \quad \varepsilon_\vartheta^\alpha = \mu_\alpha - \bar{\mu}_\alpha.$$

9.3 Information processing in the brain

In §9.2, Bayes' theorem, likelihood function, prior probability, posterior probability, Bayesian inference such as maximum likelihood estimation and MAP (maximum a posteriori) method are explained, and the variational free energy that can obtain the posterior probability approximately is described in detail. Since these are mathematical aspects, a more specific explanation of information processing in the brain will be given in this section.

Fig. 6 shows the static relationship between the environment and one-dimensional information in the brain. Expansion to multiple dimensions is easy. The purpose of information processing in the brain is to find the probability of recognition ϑ caused by the sense φ, that is, the posterior probability $p\ (\vartheta \mid \varphi)$, and it is nothing but to find the ϑ that minimizes the variational free energy. Figure 1 shows the "sensor-environment model", "recognition-sensor model", and "belief-recognition model", respectively.

$$\varphi = S + \varepsilon_\gamma, \tag{9.63a}$$

$$\varphi = g(\vartheta) + \varepsilon_z, \tag{9.63b}$$

$$\vartheta = f(\vartheta_d) + \varepsilon_w, \tag{9.63c}$$

where ε_γ, ε_z, ε_w are the errors of each model. The related Bayes' theorem is shown in Fig. 9.6.

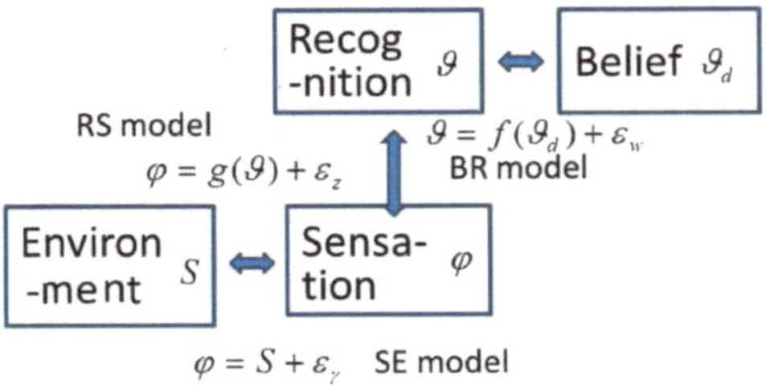

Fig. 9.5 Information flow in environment and brain

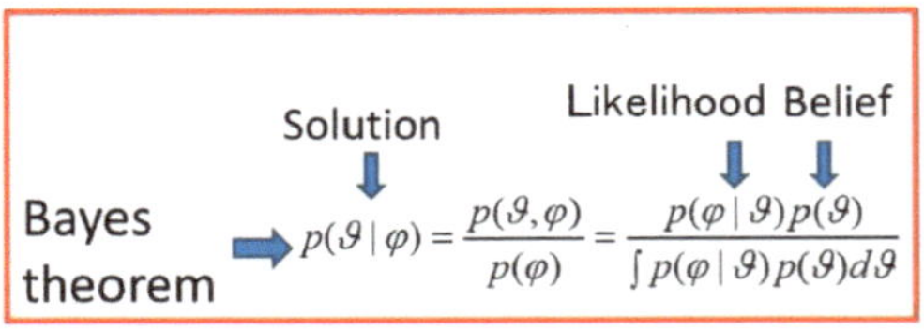

Fig. 9.6 Bayes theorem

Friston argues that in the brain, the posterior probability $p\ (\vartheta \mid \varphi)$, which is the solution, is obtained by applying the Laplace approximation (normal distribution) to the variational free energy (VFE). The static case of one variable is explained.

$$p(\varphi \mid \vartheta) = \frac{1}{\sqrt{2\pi\sigma_z}} \exp\left(-\frac{(\varphi - g(\vartheta))^2}{2\sigma_z}\right), \quad p(\vartheta) = \frac{1}{\sqrt{2\pi\sigma_w}} \exp\left(-\frac{(\vartheta - f(\vartheta_d))^2}{2\sigma_w}\right) \tag{9.64a}$$

$$p(\varphi,\vartheta)=\frac{1}{\sqrt{2\pi\sigma_z}}\frac{1}{\sqrt{2\pi\sigma_w}}\exp\left(-\frac{(\varphi-g(\vartheta))^2}{2\sigma_z}-\frac{(\vartheta-\vartheta_d)^2}{2\sigma_w}\right) \tag{9.64b}$$

σ_z and σ_w are variances. Assuming here $q(\vartheta)=\delta(\vartheta-\mu)$, we have

$$\begin{aligned}F &\equiv KL(\delta(\vartheta-\mu)\,\|\,p(\vartheta,\varphi))\\ &=\int\delta(\vartheta-\mu)\ln\frac{\delta(\vartheta-\mu)}{p(\vartheta,\varphi)}d\vartheta=\delta(0)-\ln p(\mu,\varphi)=-\ln p(\mu,\varphi).\end{aligned} \tag{9.65}$$

Since the term of δ (0) is a constant, it can be ignored in the discussion of finding the minimum value of F. In the following discussion, the energy E is defined as:

$$E[\mu]\equiv-\ln p(\mu,\varphi). \tag{9.66}$$

Substitution of equation (9.64b) into equation (9.66) gives the following equation.

$$F[\mu]=E[\mu]=\frac{1}{2}\left(\frac{(\varphi-g(\mu))^2}{\sigma_z}+\frac{(\mu-\vartheta_d)^2}{\sigma_w}+\ln(2\pi\sigma_z)+\ln(2\pi\sigma_w)\right). \tag{9.67}$$

Here, F = min corresponds to MAP method of Bayesian inference, and it is close to the form of least squares weighted by inverse of variance or accuracy. The flow of information in the brain is shown in Fig. 9.7. The minimization of variational free energy is performed by the steepest descent method:

$$(\mu,a)=\underset{(\mu,a)}{\arg\min}\,E(\mu,\varphi)=\underset{(\mu,a)}{\arg\min}\frac{1}{2}\left(\frac{(\varphi-g(\mu))^2}{\sigma_z}+\frac{(\mu-\vartheta_d)^2}{\sigma_w}+\log(\sigma_z\sigma_w)\right), \tag{9.68}$$

$$\text{Steepest Decent Method using}\ \frac{\partial E}{\partial\mu},\frac{\partial E}{\partial a}=\frac{\partial E}{\partial\varphi}\frac{d\varphi}{aa}. \tag{9.69}$$

where a is action. Action a is included in φ.

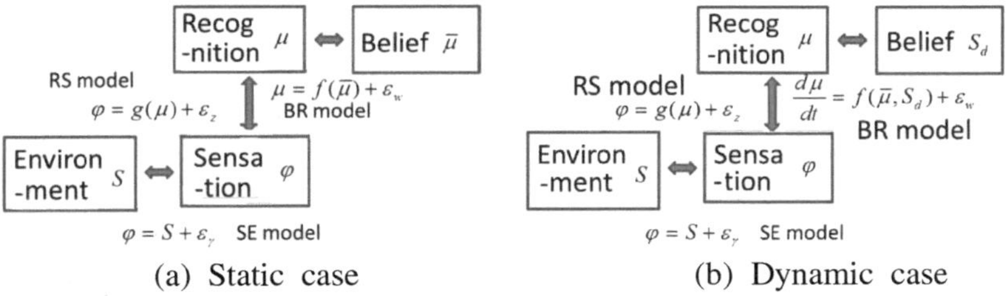

(a) Static case (b) Dynamic case

Fig. 9.7 Information flow in brain

Fig. 8 shows an example of calculation when $g(\mu) = \mu$. The recognition μ approaches the environment S (temperature T in this example; mean of sensation φ) or the belief ϑ_d (mu_bar in Fig. 7) due to the conflict between the sensation and the will. It is possible to improve the accuracy of the sensation by the action a and learn the variance in the same way to match the sensation, recognition, and belief. The recognition μ is dragged to sensation φ in Fig. 8(a). The recognition μ is intermediate of sensation φ and belief $\bar{\mu}$ in Fig. 8(b). The recognition μ is dragged to belief $\bar{\mu}$ in Fig. 8(c).

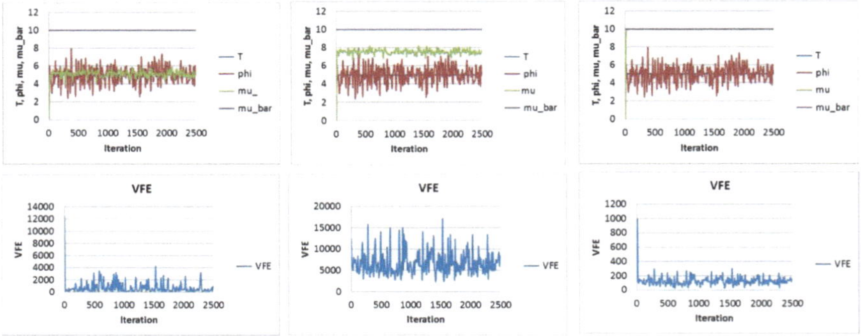

Sensation>>Belief $\sigma_z = 0.001,\ \sigma_w = 0.1$

Sensation=Belief $\sigma_z = 0.001,\ \sigma_w = 0.001$

Sensation<<Belief $\sigma_z = 0.1,\ \sigma_w = 0.001$

Fig. 9.8 Conflict of sensation and belief ($\gamma = 1.0$, $k = 0.0002$)

Consider why the result shown in Fig. 9.8 is obtained. When $g(\mu) = \mu$, E is given

$$E(\mu,\varphi) = \arg\min_{\mu} \frac{1}{2}\left(\frac{(\varphi-\mu)^2}{\sigma_z} + \frac{(\mu-\vartheta_d)^2}{\sigma_w} + \log(\sigma_z\sigma_w) \right) \tag{9.70}$$

we have

$$\frac{\partial E}{\partial \mu} = \frac{-(\varphi-\mu)}{\sigma_z} + \frac{(\mu-\vartheta_d)}{\sigma_w} = 0. \tag{9.71}$$

From this, we obtain

$$\mu = \frac{1}{\sigma_z+\sigma_w}(\sigma_w\varphi+\sigma_z\vartheta_d) \to \begin{cases} \varphi & \text{when } \sigma_z \ll \sigma_w \\ \vartheta_d & \text{when } \sigma_z \gg \sigma_w \end{cases} \tag{9.72}$$

As mentioned above, it can be seen from the mathematical formula that recognition is dragged by sensation and belief according to the given conditions.

9.4 Information processing of dynamic problems

Friston's idea of dynamic problems [4] is complicated and difficult to understand. The idea of a dynamic problem different from Friston [4] through the same problem as Buckley [2] is shown below. I think it's much easier to understand.

There is a heat source like a stove. The closer you get, the higher the temperature, and the farther you move, the lower the temperature. We want to go to the place of our favorite temperature. It is assumed that the distribution of the temperature T_i at the time t_i and the coordinate x_i in the one-dimensional space is given by the following equation:

$$T_i = \frac{T_0}{1+x_i^2}, \tag{9.73a}$$

$$\left(\frac{dT}{dx}\right)_i = -\frac{2x_i T_0}{(1+x_i^2)^2}. \tag{9.73b}$$

T_0 is the temperature at $x_i = 0$. "Environment T_i - sensory φ_i", "sensory φ_i - recognition μ_i" and "recognition μ_i - will μ_{di}" models are given by

$$\varphi_i = T_i + \varepsilon_{\gamma i}, \tag{9.74b}$$

$$\varphi_i = g(\mu_i) + \varepsilon_{zi}, \tag{9.74b}$$

$$\mu_i = \mu_{i-1} - dt(\mu_{i-1} - \mu_{d\,i-1}) + \varepsilon_{w\,i-1}. \tag{9.74c}$$

Here, $\varepsilon_{\gamma i}$, ε_{zi} and ε_{wi} are noise. According to equation (9.74c), it can be seen that the solution converges to μ_d when $t_i \to \infty$. The unknown variables in this problem are the recognition μ_i and the action x_i. The action x_i is hidden in φ_i through equations (9.73) and (9.74b). This problem can be solved by the steepest descent method using the principle of minimum variational energy. The principle of minimum variational energy in this case is that E_i is the energy, and p (μ_i, φ_i) is the joint probability distribution of μ_i and φ_i :

$$E_i[\mu_i, x_i] = -\ln p(\mu_i, \varphi_i) = \min . \tag{9.75}$$

Here, since we have from equations (9.74) and (9.75)

$$p(\mu_i, \varphi_i) = p(\varphi_i \mid \mu_i) p(\mu_i)$$
$$= \frac{1}{\sqrt{2\pi\sigma_{zi}}} \exp\left(-\frac{(\varphi_i - g(\mu_i))^2}{2\sigma_{zi}}\right) \frac{1}{\sqrt{2\pi\sigma_{wi}}} \exp\left(-\frac{(\mu_i - [\mu_{i-1} - dt(\mu_{i-1} - \mu_{d\,i-1})])^2}{2\sigma_{wi}}\right), \tag{9.76}$$

we obtain

$$E_i = -\ln p_i(\mu_i, \varphi_i) = \frac{(\varphi_i - g(\mu_i))^2}{2\sigma_{zi}} + \frac{(\mu_i - [\mu_{i-1} - dt(\mu_{i-1} - \mu_{d\,i-1})])^2}{2\sigma_{wi}}$$
$$+\frac{1}{2}\ln(2\pi\sigma_{zi}) + \frac{1}{2}\ln(2\pi\sigma_{wi}). \tag{9.77}$$

The recognition μ_i of energy E_i and the differentiation by action x_i are as follows

$$\frac{\partial E_i}{\partial \mu_i} = -\frac{\varphi_i - g(\mu_i)}{\sigma_{zi}} g'(\mu_i) + \frac{\mu_i - [\mu_{i-1} - dt(\mu_{i-1} - \mu_{d\,i-1})]}{\sigma_{wi}}, \tag{9.78a}$$

$$\frac{\partial E_i}{\partial x_i} = \frac{\partial E_i}{\partial \varphi_i}\frac{\partial \varphi_i}{\partial x_i} = -\frac{(\varphi_i - \mu_i)}{\sigma_{zi}}\left(\frac{dT}{dx}\right)_i, \tag{9.78b}$$

$$\mu_i = \mu_{i-1} - \kappa\frac{\partial E_i}{\partial \mu_i}, \quad x_i = x_{i-1} - \kappa\frac{\partial E_i}{\partial x_i}. \tag{9.79c}$$

A calculation example is shown in Fig. 9. κ is a learning step. The parameters used in the calculation ($g\ (\mu) = \mu$) are given by

$$\begin{aligned} &T_0 = 100, T_d = 4, x_0 = 2, dt = 0.01, k = 0.001, \\ &\mu_0 = 10, \sigma_\gamma = 0.1, \sigma_z = 0.1, \sigma_w = 0.01, \textit{ation start time} = 25. \end{aligned} \tag{9.80}$$

Until the action x is turned on, there is a difference between the recognition μ and the sensation φ, but when the action x is turned on at t = 25, the difference between the two disappears, and the movement starts and finally reaches the position x = 5. It reaches the desired temperature T_d = 4. It can also be seen that the variational free energy drops sharply when the action is applied. In Fig. 9 (left), T and φ overlap.

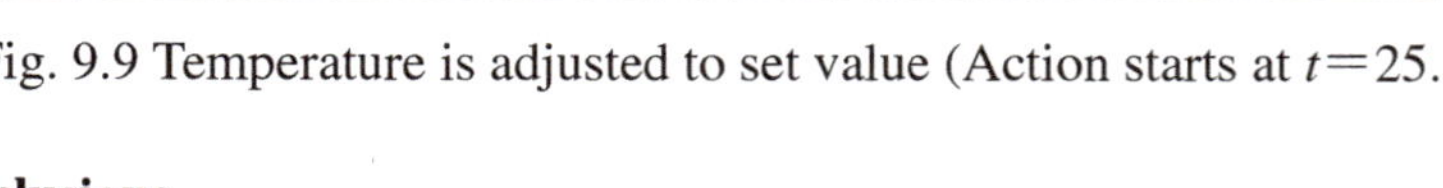

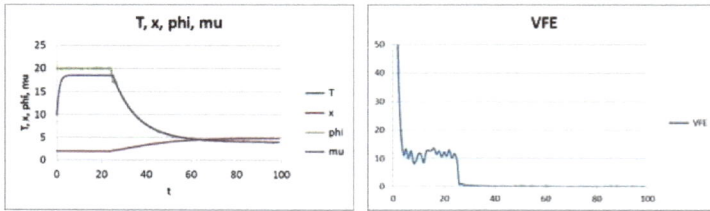

Fig. 9.9 Temperature is adjusted to set value (Action starts at t=25.)

9.5 Conclusions

We discussed the mathematics of information processing in the brain. We followed the free energy proposed by Friston [1,4]. There are excellent explanations by Buckley [2] and Bogacz [3], but they are quite difficult, so we aimed to explain it from a different point of view. The former is characterized by dynamic aspects and the latter by learning and biological implementation. We were able to give it some originality by organizing it and adding new calculations.

Friston's explanation of dynamic problems is verbose, complex and confusing. We think we were able to show a much easier way of thinking. Actually, the same calculation as Buckley [2] was performed and useful results were obtained, so we think it is a big improvement.

REFERENCES 9

[1] K. Friston, *Nature Reviews | Neuroscie*, Vol. 11, 2010.

[2] C. L. Buckley et al., *Journal of Mathematical Psychology* 81, 2017, 55–79.

[3] R. Bogacz, *Journal of Mathematical Psychology*, 76, 2017, 198-211.

[4] K. Friston, Hierarchical Models in the Brain, *PLoS Computational Biology*, Vol. 4, Issue 11, 2008, 1-24.

[5] Vase/faces – what do you see? | Optical illusions for kids, Face illusions, Optical,
https://www.pinterest.cl/pin/59743132535430394/

[6] The Rubber Hand Illusion | Horizon | BBC Studios,
https://www.youtube.com/watch?v=Qsmkgi7FgEo.

Appendix 9A Non-negativity of Kullback-Leibler divergence

The natural logarithm has the following properties:

$$\ln x \leq x-1. \tag{9A.1}$$

If the probability distribution is continuous, then if Θ is the set of all ϑ where $q(\vartheta)$ is not zero, then the following equation holds:

$$-\int_{\Theta} q \ln \frac{p}{q} d\vartheta \geq -\int_{\Theta} q\left(\frac{p}{q}-1\right) d\vartheta = -\int_{\Theta} p d\vartheta + \int_{\Theta} q d\vartheta = -\int_{\Theta} p d\vartheta + 1 \geq 0. \tag{A9.2}$$

Therefore, we obtain:

$$-\int_{\Theta} q \ln p d\vartheta \geq -\int_{\Theta} q \ln q d\vartheta. \tag{A9.3}$$

Even if 0 is added to both sides, the magnitude relationship does not change, so the following equation holds:

$$\int_{-\infty}^{\infty} q \ln p d\vartheta \leq \int_{-\infty}^{\infty} q \ln q d\vartheta. \tag{A9.4}$$

Therefore, in the case of continuous probabilities, the non-negativeness of Kullback-Leibler divergence is derived:

$$\int_{-\infty}^{\infty} q \ln \frac{p}{q} d\vartheta \leq 0 \rightarrow \int_{-\infty}^{\infty} q \ln \frac{q}{p} d\vartheta \geq 0. \tag{A9.5}$$

If the probability distribution is discrete, let I be the set of all i whose q_i is not zero.

$$-\sum_{i\in I} q_i \ln\frac{p_i}{q_i} \geq -\sum_{i\in I} q_i\left(\frac{p_i}{q_i}-1\right) = -\sum_{i\in I} p_i + \sum_{i\in I} q_i = -\sum_{i\in I} p_i + 1 \geq 0. \qquad \text{(A9.6)}$$

Therefore, we obtain:

$$-\sum_{i\in I} q_i \ln p_i \geq -\sum_{i\in I} q_i \ln q_i. \qquad \text{(A9.7)}$$

Even if 0 is added to both sides, the magnitude relationship does not change, so the following equation holds:

$$\sum_{i=1}^{n} q_i \ln p_i \leq \sum_{i=1}^{n} q_i \ln q_i. \qquad \text{(A9.8)}$$

Therefore, in the case of discrete probabilities, the non-negativeness of Kullback-Leibler divergence is also derived:

$$\sum_{i=1}^{n} q_i \ln\frac{p_i}{q_i} \leq 0 \rightarrow \sum_{i=1}^{n} q_i \ln\frac{q_i}{p_i} \geq 0. \qquad \text{(A9.9)}$$

Appendix 9B C-language code for numerical calculation of posterior probability $p(\vartheta\,|\,\varphi)$ using variational free energy

"Microsoft C/C++ Compiler Version 17.00.50727.1 for x86" and "Microsoft Linker Version 11.00.50727" were used for compile and link (in command window; cl source_file_name.c).

(1) Progrming code: PostProbNCalByVFE.c

```
// -------------------------------------------------------------------------- //
//                                                                            //
// File Name: PostProbNCal.c                          2022.04.25-2022.04.28 //
// File Name: PostProbNCalByVFE.c                     2022.04.28-2022.05.08 //
//                                                                            //
//    Numerical Cal. of Posteriro Prob.                                       //
//                                                                            //
// -------------------------------------------------------------------------- //

#include <stdio.h>
#include <stdlib.h>
#include <string.h>
#include <math.h>
```

```
#define PI      3.14159265358979323846     // 円周率の定義

// ----- function ------------------------------------------------------- //

void main();
void pushKey();

double F(double, double, int);                    // F(myu,zeta)
double M(double, double, int);                    // M(myu,zeta)
double Z(double, double, int);                    // Z(myu,zeta)

double DMmyu(double, double, int);                // DM/Dmyu(myu,zeta,m)
double DMzeta(double, double, int);               // DM/Dzeta(myu,zeta,m)
double DZmyu(double, double, int);                // DZ/Dmyu(myu,zeta,m)
double DZzeta(double, double, int);               // DZ/Dzeta(myu,zeta,m)

// ----- 変数 ----------------------------------------------------- //

char title_memo[5000];

int Ntht;                        // number of divisiob of theta
int Nphi;                        // number of divisiob of phi

int ITR;                         // number of iteration in Newton-Raphson method

double Tht;                      // -Tht <= tht <= Tht
double tht[2001];                // theta
double dtht;                     // dtheta
double Phi;                      // -Phi <= tht <= Phi
double phi[101];                 // phi
double dphi;                     // dphi

double myu;                      // mean of probability q(tht)
double zeta;                     // variance of probability q(tht)
double myu_ini;                  // initial value of myu
double zeta_ini;                 // initial value of zeta
double dmyu;                     // dmyu
double dzeta;                    // dzeta
```

```
double myub;                        // myyu_bar
double Sphi;                        // variance of liklihood functio p(phi|tht)
double Stht;                        // variance of prior probability p(tht)
double delm;                        // delm
double delz;                        // delz
double alp;                         // parameter for iteration

double myuRec[101];                 // mean of p(tht|phi);
   double zetaRec[101];                // variance of p(tht|phi);

   double p_phiBtht[20001][101];        // p(tht|phi): liklihood function
   double p_tht[20001];                 // p(tht)      : prior probability
   double p_tht_phi[2001][101];        // p(tht,phi): joint probability
   double p_phi[2001];                 // p(phi)      : marginal distribution
   double p_thtBphi[2001][101];        // p(tht|phi): posterior probability

   double p_thtBphi_VFE[2001][101];  // p(tht|phi): posterior probability by VFE

   FILE *fp_inp;                       // file pointer of input file
   FILE *fp_out;                       // file pointer of output file

   char InputDataFile[80];             // input file name
   char OutputDataFile[80];            // output file name

   char buf[5000];                     // buffer

   int prt_ctrl;                       // print control; if prt_ctrl = 1, then print

   // ---------------------------------------------------------------- //

   void main()
   {
       int i, m, itr;
       double sum;

       //// open input file for parameters
       sprintf(InputDataFile, "PostProbNCalByVFE_inp.dat");

       if ((fp_inp = fopen(InputDataFile, "r")) == NULL) {
```

```
    printf("Failed in Reading Input Data File! ... %s¥n", InputDataFile);
    exit(1);
}

//// open output file
sprintf(OutputDataFile, "PostProbNCalByVFE_out.csv");

if ((fp_out = fopen(OutputDataFile, "w")) == NULL) {
    printf("Failed in Reading Output Data File! ... %s¥n", OutputDataFile);
    exit(1);
}

//// input from file
fscanf(fp_inp, "%s", title_memo);

fscanf(fp_inp, "%s %d", buf, &Ntht);
fscanf(fp_inp, "%s %d", buf, &Nphi);

fscanf(fp_inp, "%s %d", buf, &ITR);

fscanf(fp_inp, "%s %lf", buf, &Tht);
fscanf(fp_inp, "%s %lf", buf, &Phi);

fscanf(fp_inp, "%s %lf", buf, &myub);
fscanf(fp_inp, "%s %lf", buf, &Sphi);
fscanf(fp_inp, "%s %lf", buf, &Stht);

fscanf(fp_inp, "%s %lf", buf, &myu_ini);
fscanf(fp_inp, "%s %lf", buf, &zeta_ini);

fscanf(fp_inp, "%s %lf", buf, &delm);
fscanf(fp_inp, "%s %lf", buf, &delz);
fscanf(fp_inp, "%s %lf", buf, &alp);

fscanf(fp_inp, "%s %d", buf, &prt_ctrl);
fclose(fp_inp);

//// output to display
printf("memo: %s¥n", title_memo);
```

```
printf("Ntht  = %d¥n", Ntht);
printf("Nphi  = %d¥n", Nphi);

printf("ITR  = %d¥n", ITR);

printf("Tht   = %12.6f¥n", Tht);
printf("Phi   = %12.6f¥n", Phi);

printf("myub  = %12.6f¥n", myub);
printf("Sphi  = %12.6f¥n", Sphi);
printf("Stht  = %12.6f¥n", Stht);

printf("myu_ini  = %12.6f¥n", myu_ini);
printf("zeta_ini = %12.6f¥n", zeta_ini);

printf("delm     = %12.6f¥n", delm);
printf("delz     = %12.6f¥n", delz);
printf("alp      = %12.6f¥n", alp);

printf("prt_ctrl = %d¥n", prt_ctrl);
printf("¥n");

//// output to file
fprintf(fp_out, "memo: %s¥n", title_memo);
fprintf(fp_out, "¥n");

fprintf(fp_out, "Ntht =, %d¥n", Ntht);
fprintf(fp_out, "Nphi =, %d¥n", Nphi);

fprintf(fp_out, "ITR =, %d¥n", ITR);

fprintf(fp_out, "Tht =, %12.6f¥n", Tht);
fprintf(fp_out, "Phi =, %12.6f¥n", Phi);

fprintf(fp_out, "myub =, %12.6f¥n", myub);
fprintf(fp_out, "Sphi =, %12.6f¥n", Sphi);
fprintf(fp_out, "Stht =, %12.6f¥n", Stht);
```

```
    fprintf(fp_out, "myu_ini =, %12.6f¥n", myu_ini);
    fprintf(fp_out, "zeta_ini =, %12.6f¥n", zeta_ini);

    fprintf(fp_out, "delm =, %12.6f¥n", delm);
    fprintf(fp_out, "delz =, %12.6f¥n", delz);
    fprintf(fp_out, "alp =, %12.6f¥n", alp);

    fprintf(fp_out, "prt_ctrl =, %d¥n", prt_ctrl);
    fprintf(fp_out, "¥n");

    pushKey();


    //////////////////////////////
    //                          //
    //   Analytical Solution    //
    //                          //
    //////////////////////////////

    dtht = 2.0*Tht/(Ntht+0.0);
    for (i = 0; i < Ntht; i++)
        tht[i] = -Tht+(i+0.5)*dtht;

    dphi = 2.0*Phi/(Nphi+0.0);
    for (m = 0; m < Nphi; m++)
        phi[m] = -Phi+(m+0.5)*dphi;


    // p(phi|tht): liklihood function
    for (m = 0; m < Nphi; m++)
        for (i = 0; i < Ntht; i++)
            p_phiBtht[i][m] =
1.0/sqrt(2.0*PI*Sphi)*exp(-(phi[m]-tht[i])*(phi[m]-tht[i])/2.0/Sphi);


    fprintf(fp_out, "**********¥n");
    i = Ntht/2;
    fprintf(fp_out, "i, m, tht[Ntht/2], phi[m], p_phiBtht[Ntht/2][m]¥n");
    for (m = 0; m < Nphi; m++)
```

```
    fprintf(fp_out, "%d, %d, %lf, %lf, %lf¥n",
            i, m, tht[i], phi[m], p_phiBtht[i][m]);
fprintf(fp_out, "¥n");



fprintf(fp_out, "p(phi|tht):, liklihood function¥n");
fprintf(fp_out, "phi, ");
for (i = 0; i < Ntht; i++)
    fprintf(fp_out, "tht=%6.3f, ", tht[i]);
fprintf(fp_out, "¥n");
for (m = 0; m < Nphi; m++) {
    fprintf(fp_out, "%lf, ", phi[m]);
    for (i = 0; i < Ntht; i++)
        fprintf(fp_out, "%lf, ",  p_phiBtht[i][m]);
    fprintf(fp_out, "¥n");
}
fprintf(fp_out, "¥n");

// p(tht): prior probability
fprintf(fp_out, "p(tht):, prior probability¥n");
fprintf(fp_out, " i, tht, p_tht¥n");
for (i = 0; i < Ntht; i++) {
    p_tht[i] = 1.0/sqrt(2.0*PI*Stht)*exp(-(tht[i]-myub)*(tht[i]-myub)/2.0/Stht);
    fprintf(fp_out, "%d, %lf, %lf¥n", i, tht[i], p_tht[i]);
}
fprintf(fp_out, "¥n");

// p(tht,phi): joint probability
for (i = 0; i < Ntht; i++)
    for (m = 0; m < Nphi; m++)
        p_tht_phi[i][m] = 1.0/sqrt(2.0*PI*Stht)*1.0/sqrt(2.0*PI*Sphi)
                          *exp(-(phi[m]-tht[i])*(phi[m]-tht[i])/2.0/Sphi
                              -(tht[i]-myub)*(tht[i]-myub)/2.0/Stht);

fprintf(fp_out, "p(thtCphi):, joint probability¥n");
fprintf(fp_out, "tht, ");
for (m = 0; m < Nphi; m++)
    fprintf(fp_out, "phi=%6.3f, ", phi[m]);
```

```
fprintf(fp_out, "¥n");
for (i = 0; i < Ntht; i++) {
    fprintf(fp_out, "%lf, ", tht[i]);
    for (m = 0; m < Nphi; m++)
        fprintf(fp_out, "%lf, ",  p_tht_phi[i][m]);
    fprintf(fp_out, "¥n");
}
fprintf(fp_out, "¥n");

// p(phi): marginal distribution
fprintf(fp_out, "p(phi):, marginal probability¥n");
fprintf(fp_out, "m, phi, p(phi)¥n");
for (m = 0; m < Nphi; m++) {
    p_phi[m] = 1.0/sqrt(2.0*PI*(Stht+Sphi))*exp(-(phi[m]-myub)*(phi[m]-myub)
                                  /2.0/(Stht+Sphi));
    fprintf(fp_out, "%d, %lf, %lf¥n", m, phi[m], p_phi[m]);
}
fprintf(fp_out, "¥n");

// p(tht|phi): posterior probability...analytical
for (i = 0; i < Ntht; i++)
    for (m = 0; m < Nphi; m++)
        p_thtBphi[i][m] = p_tht_phi[i][m]/p_phi[m];

fprintf(fp_out, "p(tht|phi):, posterior probabilityy...analytical¥n");
fprintf(fp_out, "tht, ");
for (m = 0; m < Nphi; m++)
    fprintf(fp_out, "phi=%6.3f, ", phi[m]);
fprintf(fp_out, "¥n");
for (i = 0; i < Ntht; i++) {
    fprintf(fp_out, "%lf, ", tht[i]);
    for (m = 0; m < Nphi; m++)
        fprintf(fp_out, "%lf, ",  p_thtBphi[i][m]);
    fprintf(fp_out, "¥n");
}
fprintf(fp_out, "¥n");

fprintf(fp_out, "p(tht|phi):, posterior probabilityy...analytical¥n");
fprintf(fp_out, "phi, ");
```

```
for (i = 0; i < Ntht; i++)
    fprintf(fp_out, "tht=%6.3f, ", tht[i]);
fprintf(fp_out, "¥n");
for (m = 0; m < Nphi; m++) {
    fprintf(fp_out, "%lf, ", phi[m]);
    for (i = 0; i < Ntht; i++)
        fprintf(fp_out, "%lf, ",  p_thtBphi[i][m]);
    fprintf(fp_out, "¥n");
}
fprintf(fp_out, "¥n");

//////////////////////////////
//                          //
// Numerical Solution by    //
// Variational Free Energy //
//                          //
//////////////////////////////

// Newton Raphson

myu = myu_ini;
zeta = zeta_ini;

for (m = 0; m < Nphi; m++) {
    if (prt_ctrl == 1) {
        fprintf(fp_out, "m =, %d, phi =, %lf¥n", m, phi[m]);
        fprintf(fp_out, "itr, dmyu, dzeta, myu, zeta, F¥n");
        fprintf(fp_out, "%d, %lf, %lf, %lf, %lf¥n", 0, 0.0, 0.0, myu, zeta);
    }
    for (itr = 1; itr <= ITR; itr++) {
        dmyu = -(M(myu, zeta, m)*DZzeta(myu, zeta, m) - Z(myu, zeta, m)*DMzeta(myu, zeta, m))
                /(DMmyu(myu, zeta, m)*DZzeta(myu, zeta, m)
                  - DZmyu(myu, zeta, m)*DMzeta(myu, zeta, m));
        dzeta = -(DMmyu(myu, zeta, m)*Z(myu, zeta, m) - DZmyu(myu, zeta, m)*M(myu, zeta, m))
                /(DMmyu(myu, zeta, m)*DZzeta(myu, zeta, m)
                  - DZmyu(myu, zeta, m)*DMzeta(myu, zeta, m));

        myu += alp*dmyu;
        zeta += alp*dzeta;
```

```
        if (prt_ctrl == 1)
            fprintf(fp_out, "%d, %lf, %lf, %lf, %lf, %lf¥n",
                    itr, dmyu, dzeta, myu, zeta, F(myu,zeta,m));
    }
    if (prt_ctrl == 1)
        fprintf(fp_out, "¥n");

    myuRec[m] = myu;
    zetaRec[m] = zeta;
}

fprintf(fp_out, "mean and variance of p(tht|phi)¥n");
fprintf(fp_out, "m, phi,myuRec, my_ex, zetaRec, zeta_ex¥n");
for (m = 0; m < Nphi; m++)
    fprintf(fp_out, "%d, %lf, %lf, %lf, %lf, %lf¥n",
            m, phi[m], myuRec[m], (Sphi*myub+Stht*phi[m])/(Stht+Sphi),
            zetaRec[m], Sphi*Stht/(Stht+Sphi));
fprintf(fp_out, "¥n");

// p(tht|phi) by VFE)
for (i = 0; i < Ntht; i++)
    for (m = 0; m < Nphi; m++)
        p_thtBphi_VFE[i][m] = 1.0/sqrt(2.0*PI*zetaRec[m])*exp(-(tht[i]-myuRec[m])
                              *(tht[i]-myuRec[m])/2.0/zetaRec[m]);

fprintf(fp_out, "p(tht|phi):, posterior probability by VFE...numerical¥n");
fprintf(fp_out, "tht, ");
for (m = 0; m < Nphi; m++)
    fprintf(fp_out, "phi=%6.3f, ", phi[m]);
fprintf(fp_out, "¥n");
for (i = 0; i < Ntht; i++) {
    fprintf(fp_out, "%lf, ", tht[i]);
    for (m = 0; m < Nphi; m++)
        fprintf(fp_out, "%lf, ",  p_thtBphi_VFE[i][m]);
    fprintf(fp_out, "¥n");
}
fprintf(fp_out, "¥n");
```

```
    fclose(fp_out);

    pushKey();
}

// ------------------------------------------------------------ //

 void pushKey()
{
    printf("¥n       Push Return Key! ");
    getchar();
    getchar();
}

// ------------------------------------------------------------ //

double F(double myu, double zeta, int m)
{
    int i;
    double sum;

    sum = 0.0;
    for (i = 0; i < Ntht; i++)
        sum += exp(-(tht[i]-myu)*(tht[i]-myu)/2.0/zeta)*(phi[m]-tht[i])
                    *(phi[m]-tht[i])/2.0/Sphi;
    sum *= dtht;
    sum *= 1.0/sqrt(2.0*PI*zeta);

    return (-0.5*log(zeta)+0.5*log(Sphi*Stht)+0.5*log(2.0*PI)
            +(myu-myub)*(myu-myub)/2.0/Stht)
            + zeta*(-1.0/2.0/zeta+1.0/2.0/Stht) + sum;
}

// ------------------------------------------------------------ //

double M(double myu, double zeta, int m)
{
    return (F(myu+delm,zeta,m)-F(myu,zeta,m))/delm;
}
```

```
// ------------------------------------------------------------ //

double Z(double myu, double zeta, int m)
{
    return (F(myu,zeta+delz,m)-F(myu,zeta,m))/delz;
}

// ------------------------------------------------------------ //

double DMmyu(double myu, double zeta, int m)
{
    return (M(myu+delm,zeta,m)-M(myu,zeta,m))/delm;
}

// ------------------------------------------------------------ //

double DMzeta(double myu, double zeta, int m)
{
    return (M(myu,zeta+delz,m)-M(myu,zeta,m))/delz;
}

// ------------------------------------------------------------ //

double DZmyu(double myu, double zeta, int m)
{
    return (Z(myu+delm,zeta,m)-Z(myu,zeta,m))/delm;
}

// ------------------------------------------------------------ //

double DZzeta(double myu, double zeta, int m)
{
    return (Z(myu,zeta+delz,m)-Z(myu,zeta,m))/delz;
}

// ------------------------------------------------------------ //
```

(2) Input file

```
2022.05.08

Ntht            101
Nphi            41

ITR             20

Tht             5.0
Phi             4.0

myub            0.0
Sphi            1.5
Stht            2.5

myu_ini        -2.5
zeta_ini        1.0

delm            0.000001
delz            0.000001
alp             0.25

prt_ctrl       0
```

Appendix 9C C-language code for dynamic control problem using variational free energy

"Microsoft C/C++ Compiler Version 17.00.50727.1 for x86" and "Microsoft Linker Version 11.00.50727" were used for compile and link (in command window; cl source_file_name.c).

(1) Programing code: BuckleyXHINewDyn_Walk.c

```
/* ------------------------------------------------------------------ */
/*                                                                    */
/* File Name: BuckleyX.c                   2021.06.23-2021.06.26 */
/* File Name: BuckleyXHI.c                 2021.06.26-2021.07.02 */
/* File Name: BuckleyXHINewDyn.c           2021.07.26-2021.08.12 */
/* File Name: BuckleyXHINewDyn_Walk.c      2021.08.12-2022.05.04 */
/*                                                                    */
/*    Variational Free Energy                                         */
/*      rho: phai                                                     */
```

```
/*                                                                   */
/* ----------------------------------------------------------------- */

#include <stdio.h>
#include <stdlib.h>
#include <string.h>
#include <math.h>

#define PI      3.14159265

void main();
void pushKey();

double sgn(double);          // 1 if x>0; -1 if x<0; 0 if x=0

void func(double x,double y[],double f[]);

double Uniform( void );                         // Uniform random number in [0,1]
double rand_normal(double,double);              // Normal random number

double normal(double,double,double);            // Normaldistribution

void max_RProb();                               // Maximum value

/* ----------------------------------------------------------------- */

double simTime;                         // simulaton time
double dt;                              // time step
long N;                                 // N =simTime/dt
double time[500001];                    // time
int action;
double Td;                              //desired temperature
double actionTime;                      // action start

double phi[500001];                     // phi

double sgm_z;                           // sgm

double sgm_w;                           // sgm
```

```
double mu[500001];                    // brain state variable

double zgp[500001];                   // Sensory noise in the generative process

double gam;                           // Sensory noise scale

double a[500001];                     // action variable

double x[500001];                     // position
double T0;                            // temperature at x=0
double T[500001];                     // Temperature
double Tx[500001];                    // dT/dx

double eps_z;                         // epsz
double eps_w;                         // epsz

double VFE[500001];                   // variational free energy

double k;                             // Gradient descent learning parameters for inference

int action_bgn;                       // Action begin

int prt_skip;                         // Adjustment of sensory input

double AMAT[1001][2001];              // matrix

FILE *fp_inp;                         // pointer of input file
FILE *fp_out;                         // pointer of output file

char InputDataFile[80];               // input file name
char OutputDataFile[80];              // output file name

char buf[5000];

/* ------------------------------------------------------------ */

// A Simple Bayesian Thermostat
// The f ree energy pr inc iple for action and perception : A mathematical review , Journal
of Mathematical Psychology
```

```
void main()
{
    long i, j, n;

    // Input file
    sprintf(InputDataFile, "BuckleyXHINewDyn_Walk_inp.dat");

    if ((fp_inp = fopen(InputDataFile, "r")) == NULL) {
        printf("Failed in Reading Input Data File! ... %s¥n", InputDataFile);
        exit(1);
    }

    // Output file
    sprintf(OutputDataFile, "BuckleyXHINewDyn_Walk_out.csv");

    if ((fp_out = fopen(OutputDataFile, "w")) == NULL) {
        printf("Failed in Reading Output Data File! ... %s¥n", OutputDataFile);
        exit(1);
    }


    fscanf(fp_inp, "%s %lf", buf, &T0);
    fscanf(fp_inp, "%s %lf", buf, &Td);
    fscanf(fp_inp, "%s %lf", buf, &x[1]);
    fscanf(fp_inp, "%s %lf", buf, &simTime);
    fscanf(fp_inp, "%s %lf", buf, &dt);

    // Gradient descent learning parameters
    fscanf(fp_inp, "%s %lf", buf, &k);

    // Initialise brain s tate variables
    fscanf(fp_inp, "%s %lf", buf, &mu[1]);

    // sensory noise scales
    fscanf(fp_inp, "%s %lf", buf, &gam);

    // sensory variances
```

```
fscanf(fp_inp, "%s %lf", buf, &sgm_z);

// hidden variances
fscanf(fp_inp, "%s %lf", buf, &sgm_w);

// adjustment
fscanf(fp_inp, "%s %d", buf, &action_bgn);
fscanf(fp_inp, "%s %d", buf, &prt_skip);



printf("T0            = %12.6f¥n", T0);
printf("Td            = %12.6f¥n", Td);
printf("x[1]          = %12.6f¥n", x[1]);
printf("simTime       = %12.6f¥n", simTime);
printf("dt            = %12.6f¥n", dt);

// Gradient descent learning parameters
printf("k             = %12.6f¥n", k);

// Initialise brain s tate var iables
printf("mu[1]       = %12.6f¥n", mu[1]);

// sensory noise scales
printf("gam         = %12.6f¥n", gam);

// sensory variances
printf("sgm_z     = %12.6f¥n", sgm_z);

// hidden variances
printf("sgm_w     = %12.6f¥n", sgm_w);

// adjustment
printf("action_bgn  = %d¥n", action_bgn);
printf("prt_skip    = %d¥n", prt_skip);



fprintf(fp_out, "T0 =, %12.6f¥n", T0);
fprintf(fp_out, "Td =, %12.6f¥n", Td);
fprintf(fp_out, "X[1] =, %12.6f¥n", x[1]);
```

```
fprintf(fp_out, "simTime =, %12.6f¥n", simTime);
fprintf(fp_out, "dt =, %12.6f¥n", dt);

fprintf(fp_out, "Gradient descent learning parameters¥n");
fprintf(fp_out, "k =, %12.6f¥n", k);

fprintf(fp_out, "Initialise brain s tate var iables¥n");
fprintf(fp_out, "mu[1] =, %12.6f¥n", mu[1]);

fprintf(fp_out, "sensory noise scales¥n");
fprintf(fp_out, "gam =, %12.6f¥n", gam);

fprintf(fp_out, "sensory variances¥n");
fprintf(fp_out, "sgm_z =, %12.6f¥n", sgm_z);

fprintf(fp_out, "hidden variances¥n");
fprintf(fp_out, "sgm_w =, %12.6f¥n", sgm_w);

fprintf(fp_out, "adjustment¥n");
fprintf(fp_out, "action_bgn =, %d¥n", action_bgn);
fprintf(fp_out, "prt_skip =, %d¥n", prt_skip);
fprintf(fp_out, "¥n");

pushKey();

// simulation parameters

N =simTime/dt;
time[1] = 0.0;
for (i = 2; i <= N; i++)
    time[i] = time[i-1]+dt;

//The time that action onsets
actionTime =simTime /4;

//Sensory noise in the generative process
for (i = 1; i <= N; i++)
    zgp[i] = rand_normal(0.0,1.0)*gam;
```

```
// Initialise generative process
T[1] = T0 / (x[1]*x[1]+1.0);
Tx[1]= -2.0*T0*x[1]/(x[1]*x[1]+1.0)/(x[1]*x[1]+1.0);

// Initialise sensory input
phi[1] = T[1] ;

// Initialise error terms
eps_z = phi[1]-mu[1];
eps_w = mu[1]-mu[0]-dt*(-mu[0]+Td);

// Initialise Var iat ional Energy
VFE[1] = 1/sgm_z*eps_z*eps_z/2.0 + 1/sgm_w*eps_w * eps_w/2.0 + 1/2.0*log(sgm_w*sgm_z);


fprintf(fp_out, "i, time[i], T[i], x[i], phi[i], mu[i], VFE[i]¥n");
for (i =2; i <= N; i++) {

    phi[i-1] = T[i-1] + zgp[i-1]; //calclaute sensory input

    if ((i-2) % prt_skip == 0)
        fprintf(fp_out, "%d, %12.6f, %12.6f, %12.6f, %12.6f, %12.6f, %12.6f¥n",
                i-1, time[i-1], T[i-1], x[i-1], phi[i-1], mu[i-1], VFE[i-1]);

    //The generative model ( i . e . the agents brain )
    eps_z = phi[i-1]-mu[i-1];     // error terms: phi = mu+epsz

    eps_w = mu[i-1]-mu[i-2]-dt*(-mu[i-2]+Td);      // dmyu/dt = -myu+Td+epsw

    // Inference
    mu[i] = mu[i-1] + (-k*(-eps_z /sgm_z + eps_w /sgm_w));

    if ( time [i] > action_bgn )
        x[i] = x[i-1] + (-k*Tx [i-1]*eps_z/sgm_z);    //action
    else
        x[i] = 2.0;

    //The generative process ( i.e. the real world )
    T[i] = T0 / (x[i]*x[i]+1);
```

```
        Tx[i] = -2.0*T0*x[i]/(x[i]*x[i]+1)/(x[i]*x[i]+1);

        VFE[i] = 1/sgm_z*eps_z*eps_z/2.0 + 1/sgm_w*eps_w*eps_w/2.0 + 1/2* log(sgm_w*sgm_z);

    }
    fprintf(fp_out, "¥n");



    fclose(fp_inp);
    fclose(fp_out);

    pushKey();
}

/* ------------------------------------------------------------ */

void pushKey()
{
    printf("¥n        Push Return Key! ");
    getchar();
    getchar();
}

/* ------------------------------------------------------------ */

/* ------------------------------------------------------------ */

double sgn(double x)
{
    if (x > 0)
        return 1.0;
    else if (x < 0)
        return -1.0;
    else
        return 0.0;
}

/* ------------------------------------------------------------ */
```

```
void func(double x, double y[], double f[])
{
    f[0] = y[1];
    f[1] = 1.0 - y[0] -2.0*y[1];
}

// ------------------------------------------------------------ //

double rand_normal( double myu, double sgm )
{
  double z=sqrt( -2.0*log(Uniform()) ) * sin( 2.0*PI*Uniform() );
  return myu + sgm*z;
 }

// ------------------------------------------------------------ //

double Uniform( void )
{
  static int x=10;
  int a=1103515245,b=12345,c=2147483647;
  x = (a*x + b)&c;

  return ((double)x+1.0) / ((double)c+2.0);
}

// ------------------------------------------------------------ //

double normal(double x, double myu, double sgm)             // Normal distribution
{
    return 1.0/sqrt(2.0*PI*sgm*sgm)*exp(-(x-myu)*(x-myu)/2.0/sgm/sgm);
}

// ------------------------------------------------------------ //
```

(2) Input file: BuckleyXHINewDyn_Walk_inp.dat

```
T0              100.0
Td              4.0
X[1]            2.0
```

simTime	100.0
dt	0.01
k	0.001
mu_0[1]	10.0
gam_0	0.1
sigma_z0	0.1
sigma_w0	0.01
action_bgn	25
prt_skip	100

www.ingramcontent.com/pod-product-compliance
Lightning Source LLC
Chambersburg PA
CBHW041304210326
41598CB00005B/18